Liz Ashworth is a food writer and food product developer, with a particular interest in using local produce. The author of a pioneering series of cookery books for beginners of all ages, she writes food columns in various publications, and coordinates the food programme in the annual Orkney International Science Festival. She is also the author of *The Chain Bridge Honey Bible* and *The William Shearer Tattie Bible*.

The Barony Mill

THE
Book of Bere
Orkney's Ancient Grain

LIZ ASHWORTH

BIRLINN

First published in 2017 by
Birlinn Ltd
West Newington House
10 Newington Road
Edinburgh
EH9 1QS

British Library Cataloguing in Publication
Data. A catalogue record for this book is
available from the British Library.

ISBN: 978 1 78027 485 0

Set in Minion Pro at Birlinn

Printed and bound in Great Britain
by Latimer Trend, Plymouth

Contents

The Birsay Heritage Trust ix

Foreword xi

The Story of Bere
Orkney's Ancient Grain

What is bere? 1

    Where did bere come from? • How did bere travel to Orkney? • Pictish Orkney • The Norse on Orkney • St Magnus Cathedral • Scottish rule • Exporting bere • The kelp industry

On the land 13

    The planting • The hairst • Harvesting tradition and folklore • Mechanising agriculture • Threshing • Grinding the grain

Water and wind-powered mills come to Orkney 27

    The story of Barony Mill • Millstones • The restoration of the Barony Mill • Milling bere

The Future: The Birsay Heritage Trust 38

Research into bere 39

Why eat Orkney beremeal? 40

Time-honoured preparations of bere: brewing and distilling 41

An old Orkney kitchen 44

Recipes

Kale and knockit corn   46
Broths and soups with knockit corn   46
Burstin   47
Hints and tips on cooking with beremeal   47
Baking with beremeal   48

Bannocks

Early bere bannocks   49
Beremeal bannocks (unleavened)   49
Beremeal bannocks (unleavened, wafer-thin)   50
Auntie Betty's bere bannocks   52
Marion Whitelaw's bere bannock   53
Barbara Hewison's Westray bere bunnos   54
Davy Bain's oven-baked bere bread   55
'Bere-izza'   56
The Postie's beremeal bannocks   57
Alan Bichan's bere bannocks   58
Bere bannocks from the Creel Restaurant   **59**
The Birdy Man's bere bannocks   60
Margaret Phillips' bere bannocks   61
Lily Russell's bere bannock   61
Maureen Flett's microwave bere bannock   62

Scones and Pancakes

Victoria's beremeal blinis   64
Salt-fish blinis with nori and crab   65
Norn beremeal drop scones   66
Beremeal tattie scones   67
Joan's beremeal flatbreads
Kirsty Aim's beremeal and cheese scones   69

Seeded walnut beremeal 'bannock-scone'   70
Margaret's bacon and onion muffins   71
Bakehouse bere and raisin scones   72

Breads

Bakehouse beremeal sourdough bread   74
Beremeal and spelt baguette   76
Beremeal treacle soda bread   77

Biscuits

Beremeal and tattie crackerbread   78
Norn beremeal crackers   78
Seeded bere and oat crispbread   79
Margaret Phillips' beremeal water biscuits   80
Beremeal oatcakes   81
Beremeal and oat digestives   82
Beremeal and spelt shortbread   83
Honey crunchies   84
Ginger almond Florentines   85

Fruit Loaves

Beremeal Earl Grey tea bread   86
Fruity beremeal banana loaf cake   88
Seeded orange and date chai tea loaf   89
Beremeal, banana, chocolate and walnut loaf   90

Cakes

Highland 'Park'-in   92
Ginger 'bere' meal cake   93
Carol Wilson's beremeal fruit cake   94
Caron's cider apple cake   95

Beremeal 'Black Forest' brownies   96
Chocolate and beetroot beremeal brownies   97
Chewy date and apple bere bars   99
Fairtrade banana beremeal muffins   100
Fairtrade beremeal very 'ginger'bread   101

Savoury Dishes

Yorkney puddings   102
Spinach and hummus roulade   103
Rosanna's beremeal pasta   105
Christopher Trotter's bere 'n' carrot fritters   106
JACKS' 'berely' battered fish   107
Scottish sea trout in beremeal   108
Hand-raised North Ronaldsay mutton pie with
    sugar kelp jelly   109
Westray Wife quiche   111
Salmon quiche   112
Leek and mushroom quiche   114

Puddings

Foveran beremeal chocolate fondants   116
Bere bannock and whisky ice cream   117
Chocolate orange tart   119
Toffee apple pudding   120
Cranachan Orkney style   121

Brewing

Home-brewed ale   122

Acknowledgements   125
For Further Reading   127
Where and How to Buy Bere and Beremeal   129

The Birsay Heritage Trust

The Birsay Heritage Trust was founded in 1997 to restore the operation of the Barony Mill, which is now run by the Trust. The Mill produces stoneground beremeal and oatmeal during the winter months and is open to the public from May till September. A popular tourist attraction, it is visited by people from all over the world interested in seeing the working of a traditional water-powered mill.

It was never the aim to restrict the scope of the Trust to the Mill. A new Trust structure provides room to address wider aspects of Birsay's heritage, which is an important opportunity to preserve historic sites before they and living memories disappear and are lost for ever.

You can participate in and learn more about 5,000 years of history and tradition in Birsay by joining the Trust. New members are welcome from anywhere in the world, details on the website, *www.birsay.org.uk*.

The Birsay Heritage Trust is a recognised Scottish Charity (SCO 27642).

Barony Mills is a member of the Wind and Water-mill Section, Society for the Protection of Ancient Buildings.

A beremeal stone quarried from millstone grit

Foreword

Within this humble grain there is hidden a story longer-lasting and more action-packed than many an historic drama! But the story has not ended, for now is the beginning of a new chapter. Bere still grows in the clean sea air and fertile soil of Orkney, a living legacy waiting to be rediscovered.

In 2008, local baker Paul Groundwater and I began baking to 'investigate the potential of locally grown grain'. We started early one morning, and what a time we had! The moment we opened the bags of beremeal, it smelt better than good, and we found that with such special grain very little else was needed. Every recipe we mixed and baked came out better than the last, proving that beremeal certainly does have hidden potential.

And furthermore, beremeal is without doubt nutritionally valuable. It contains significant amounts of important nutrients, including beta glucans, fat, sugar, protein and a wide range of minerals and vitamins. The girdle-baked bere bannock is one of the oldest and healthiest of traditional daily breads. Could this be replaced as a wholesome fast food for today, a microwave bannock ready in two minutes – could the past be baked for a healthy future?

'The incorporation of foods derived from bere flours into the diet could have a beneficial impact on overall dietary quality.' (*The nutritional properties of flours derived from Orkney-grown bere*, British Nutrition Foundation, 2006.)

Bere is the descendant of a barley with impeccable celebrity endorsement. It was eaten by Cleopatra, Tutankhamun and the Queen of Sheba. Roman gladiators were known as '*hordearii*' or 'barley men' because their strong physique was attributed to a barley-based vegetarian diet. The grain which sustained countless generations of Orcadians, it was exported extensively, and caused many of them to sail the world. The bere trade helped to finance the building of St Magnus Cathedral and fed the workers who built it and visitors alike.

Growing, milling and baking bere was central to the livelihood of many families, including my own. I have discovered that my great-great-great-great-grandfather, Thomas Warren, was involved in growing bere and owned a windmill which milled it into beremeal. This was supplied to the bakery in which he was a partner, Warren and Cumming, at 16 Albert Street, Kirkwall, a short walk from where I first baked with Paul Groundwater some 200 years later in 2008. (There is a receipt handwritten and signed by Thomas Warren for the sails of a windmill, 12th October 1807 in Kirkwall Library Archives, ref. D2/19/17/21.)

The bakery remained in the Warren family until Thomas's great-grandson, also called Thomas, died in 1884, when it became Cumming and Spence, who relocated to larger premises opposite. The original Warren and Cumming bakery is now Trenabies Tearooms.

The recipes in this book are graded according to ease of making with grains of bere, one bere being simple, two bere for intermediate, and three bere requiring more advanced skills. They are testament to the quality and uses of bere and to the inventiveness of those who kindly provided recipes. Some of these have been handed down through families for generations, while others are more modern and some have been specially created for this book. All have been tried and well and truly tasted, thanks to family, friends and volunteers!

There is much to learn from the story of bere.

Corncrake, an endangered species rarely seen, in a field of bere

Stalks of bere

THE STORY OF BERE
ORKNEY'S ANCIENT GRAIN

What is bere?

Bere is a form of ancient six-row spring barley which has been grown on Orkney for thousands of years. The Scots called barley 'bere', 'bear' or 'beir', perhaps the Scottish pronunciation of a similar Old English name, 'boere'. Medieval place-names like Bearfold and Bearsden on the Scottish mainland are a reminder of its former historic importance there also. The Latin name for bere is '*Hordeum vulgare*'. It is one of the oldest cultivated 'landraces' which means it is an ancient pre-hybridised variety of barley that has evolved as nature intended, adjusting to a particular area's soil and weather conditions and through being grown from seed saved from better-yielding crops by generations of farmers. On the North Isles, bere grows rapidly during the longer daylight hours of spring and summer. Known as 90-day barley, it is sown later and harvested earlier than other similar crops.

Beremeal is the wholesome stoneground flour derived from milling bere. The cream-coloured meal has a distinctive earthy, nutty flavour with baking qualities resembling those of fine rye flour, but with a different, more rounded depth of flavour and texture.

Where did bere come from?

In 10,000 BC a lifestyle change took place in the 'fertile crescent' area between the rivers Tigris and Euphrates, historically known as Mesopotamia. The first Neolithic farming families revolutionised their way of living by the innovation of making the land 'work for them' by domesticating wild cereal crops. Early archaeological evidence of wild barley dated 8,500 BC has been found at the southern end of the Sea of Galilee and it is from such a wild barley strain (*Hordeum spontaneum*) that domesticated barley (*Hordeum vulgare*) developed.

Two and six-rowed barley differ in how kernels form on the plant head. In six-row varieties a genetic mutation causes extra grain to grow between the parallel central rows, and its greater protein content renders it valuable as a food crop for man and beast alike. This is the barley that Orcadians call 'bere'. Its more recognisable two-row cousin contains more carbohydrate and is better suited for malting to produce beer and whisky.

Following the last Ice Age, temperatures rose and the environment changed; with the adoption of agriculture the populations of the near East grew, prompting Neolithic farmers to migrate in search of fertile land. In this way, the cultivation of grain became more widespread as they moved to new areas. Farmed extensively in Egypt and Palestine, barley is mentioned in the Bible more frequently than any other grain. In Egypt six-row barley also played a significant role in religious rites and can be seen depicted in carvings on tombs and monuments. Historically, barley was appreciated for its medicinal qualities and ability to impart strength and energy, and these benefits are extolled in many religious works, including the *Torah*, the *Koran*, Hindu and Buddhist texts.

For centuries, a similar six-row barley has been grown in Tibet where it is ground into a meal resembling bere, called 'tsampa', which is mixed to a paste with yak-butter tea and eaten rolled into small balls.

How did bere travel to Orkney?

Cultivation of barley gradually spread north to Russia and Scandinavia and west through Europe to Britain. New advanced scientific techniques of genetic and morphometric analysis will help to reveal its route northwards. The fertile soils of Orkney provided an ideal environment for the first Neolithic farming families and the indigenous 'hunter-gatherer' inhabitants of these islands were absorbed into a changed community in which there was a fusion of farming and hunter-gathering. Farming thrived, so people began to settle into larger communities and built the great stone monuments that dominate the landscape of Orkney even today.

Evidence of 'naked' and hulled six-row barley has been discovered on several excavated Neolithic settlements dating as far back as 3,600BC to 2,700BC. These include Knap of Howar on Papa Westray, Pool and Toft Ness on Sanday, Barnhouse in Stenness and, in the Ha'breck area on the island of Wyre, a settlement in the period 3,300 to 3,100BC, pre-dating Skara Brae.

We discovered that house 3 was not any old farmstead. The first (and earliest) phase of use in its northern part was particularly interesting: spread across the floor were several thick layers of charred material, 70mm deep in places, comprising tens of thousands of barley grains. We had found one of the largest assemblages of Neolithic cereal in Scotland (currently under analysis by Rosie Bishop as part of her doctoral research at the University of Durham). The intensity of grain-production evidence in the house is so unusual it leads us to suspect that the building may have had a more agricultural function—a granary perhaps.

Current Archaeology, No 268, June 2012.

The inedible outer covering of what is called naked or hull-less barley is loosely attached to the kernel and falls off easily compared with the tougher protective outer skin of hulled barley. Naked barley was gradually replaced by hardier hulled grain which withstood cold wet weather and stored better if left in the ear; an important consideration in long dark Orkney winters!

Pictish Orkney

By the late Iron Age, Orkney had become part of the kingdom of the Picts. Research shows that they were living in communities based around fishing and farming, growing barley on the fertile infield and growing oats and grazing animals in the outfield. It is not certain what happened to the Picts as time went on, but as the archaeologist Caroline Wickham-Jones observes: 'Archaeological evidence shows discontinuity in settlement suggesting replacement of population, and genetic analysis indicates that Pictish women were absorbed into Norse settlements and the men sold as slaves.'

The Norse on Orkney

Orkney's fertile land and unique position on lucrative trading routes caught the eye of the Norwegian King Harald Fairhair, who reigned from 872 to 930, and he set about establishing Norse rule over the islands. By the ninth century Orkney was under Norwegian control, governed by Earls appointed by and answerable to the Norwegian crown. The majority of Norse settlements were built on good farmland with safe access to the sea, in stark contrast to the thinner soil and steep-sided fjords of Norway. There is no change in the midden waste from that period, which suggests they ate a similar diet to the native Orcadians, and certainly Norse bakers would have recognised the grain bere, because it was like a grain from home, a crop they called 'bygg'. Their sea voyages, whether to 'raid or trade', required provisions, and what better foodstuff than nutritious home-grown wholegrain bere, which was easily stored and quick to prepare and eat.

Under Norwegian Udal law, a tax called 'skat' was levied on the landowners and tenants of farming communities which were known as 'Skatland'. It was usually payable to the crown in kind, in the form of butter, malted bere, meat and poultry, or very occasionally money. To be assessed for this tax, Skatland was divided into small farming communities called 'Urislands' made up of crofts called 'pennylands'.

Udal law also decreed that family-owned land was divided by inheritance. Called the 'tounmal', this land was a strip of infield land held in perpetuity by the tenant outwith the community runrig system. It was situated near the owner's dwelling and tended like a garden. Here bere grew abundantly in soil fertilised by droppings from tethered stock and this crop was subject to even more tax! In contrast Shetland's untended 'tounmal' were tax-free.

Norse society was highly structured; earls, warriors, farmers and slaves were each allocated a work programme, some required to do more than others. Tenants of an earl's farms or 'bordlands' were exempt from skat tax but were not let off the hook because a similar tax, 'wattle', obliged them to provide food for the earl's household instead. They could also be compelled to provide a billet for the earl's men.

Earl Thorfinn the Mighty held court at Birsay where he built his Christchurch (the exact dates are difficult to find because of the few records kept then) and appointed a bishop in 1035. He was grandfather to St Magnus, whose martyrdom at the hands of his cousin Hakon had a profound and lasting effect on Orkney.

St Magnus' legacy, Kirkwall's magnificent Cathedral, plays an important part in the story of bere.

St Magnus Cathedral

St Magnus Cathedral was founded in 1137 to honour Magnus by his nephew, Earl Rognvald, who immediately began a long-term fundraising campaign to support the project. The first stage of building was financed by selling Orkney landowners the udal rights to their estates, and Earl Rognvald's next measures were taken to ensure a steady income stream for the work. In his youth Rognvald had sailed widely on trading missions, so he used the knowledge gained from this experience to develop Kirkwall into an international trading centre. As a result, in Professor Bo Almqvist's words, 'It would not be inaccurate to say that Orkney played a role in the North Sea similar to that played by Venice and other mighty Italian republics in the Mediterranean.'

Building work was not continuous, and all in all it took three centuries to complete St Magnus Cathedral. At that time, bere was also a staple grain in Europe, so craftsmen imported to work on the Cathedral and the constant stream of pilgrims and visitors who came to watch proceedings would have been familiar with Orkney's daily bread. Generations of craftsmen, labourers and pilgrims to be fed equals many extra acres of bere to be grown, milled and baked!

Each year merchants came from Bremen, Lubeck and Hamburg to Orkney to exchange goods for island produce. They brought fishing hooks, lines, nets, brandy, mead, strong beer, rye and wheat, salt, peas and fruits; also textiles like coarse linen and muslin. In return they acquired dried fish, animal hides, knitted and woven goods, fish oil, butter and mutton. The most sought-after product, however, was bere. Exported extensively to Norway, Germany and the Low Countries, bere was used for food, brewed into ale and valued for its religious significance. The 'bere boom', however, did not last. Norway was greatly affected by the Black Death and the country's subsequent union with Denmark, which led to hefty import taxes and dominance of trade routes by the German Hanseatic League. Another chapter was about to unfold.

Scottish rule

In July 1469, James III of Scotland married Margaret of Denmark at Holyrood Abbey in Edinburgh. Her father, King Christian I, being short of money, forfeited Orkney and Shetland to the Scottish crown in lieu of a dowry. The transfer of sovereignty was completed in 1472, by which time James had bought over the Norse earldoms in the islands, enabling him to collect revenue from them.

Just as in modern times there is 'nothing surer than taxes', so 'skat' became 'tack'. The King appointed a governor or 'tacksman' authorised to collect tack in return for a fee paid in money and kind to the Crown. The integrity of the appointee was important, because an able man like Lord Henry Sinclair could and did do much for Orkney's ailing agriculture, whereas others were unfortunately less scrupulous. Those liable to pay tack noticed little change in the procedure; not so the 'sub-tacksmen' who collected the levy. Numerous small payments in kind which were received in Orcadian measures such as 'lasts', 'meils' and 'settens' had then to be converted into Scottish units like 'pecks', 'bushels' and 'chalders' . . . without the aid of a calculator!

Exporting bere

Grain was transported in handwoven straw baskets called 'cassies'. A cassie is flat-bottomed and barrel-shaped, closed at the top by a drawstring of straw rope. Such was the demand for these baskets that straw-weaving became an important part of the agricultural economy. Days of bad weather were spent in the barn twisting straw rope and cord to make cassies and harnesses for pack horses. The cassies were gradually replaced during the early seventeenth century by less bulky 'meal pocks' made of 'harne', a coarse linen.

To measure grain density, a small round tub, a bushel measure, was filled to the brim with grain and a wooden stick drawn across to level the contents. The total weight less that of the measure provided the density of the grain, which varied depending on the harvest. A good crop of bere weighed in at 48lb (21.8kg) to 50lb (22.6kg) per bushel, a poor-quality lighter grain as little as 40lb (18.1kg).

Exporters faced the challenge of getting the crop to market while avoiding delays caused by storms at sea, competing with other suppliers, having to pay import tolls and duties and obtaining lower prices in times of surplus. However, there were still profits to be made, because years of poor harvest and war in importing countries created a demand for Orkney bere and subsequent high prices. Records from 1780 list exports to Lisbon in Portugal and Cadiz in Spain. By 1790 up to one third of Orkney's grain went south. The north isles of Westray and Sanday grew bere of high quality.

To quote the bible of Westray housing: 'There is a fine big store at the edge of Broughton known as Cornist which was a grain store for the Traills of Brough; the deep water meant packet boats could come well in when they came to load the grain to go south.' (*Fae Quoy tae Castle: The Buildings of Westray: An Orkney Island's Snapshot in Time*, Westray Buildings Preservation Trust, 2002.)

Bere destined for export was delivered to the harbour, weighed on a crane-like device called a 'pundler', decanted into bags of about 1cwt (50kg) and then loaded into the ship's hold. The King's tack was delivered in this way to Leith where bere was malted and exported on royal ships. By the nineteenth century, ships from Leith circumnavigated the globe carrying goods for commerce. Many Orcadian seamen relocated to Leith and sailed the world. My great grandfather, William Cumming, was one of them.

The kelp industry

In the eighteenth century, kelp seaweed gathered by generations of Orcadians for use as fertiliser became a money-spinner. Instead of agricultural use, it was burnt to extract the potash and soda it contained to supply the thriving glass and soap-making industries. For fifty years, between 1780 and 1830, the kelp industry did much to deflect the lairds' focus from the land. Making profits in excess of £22,500 per annum, three times that possible from farming, they were able to spend money amassing land and many bought or built property in Kirkwall. Workers were poorly paid in comparison, but what they did earn allowed them to buy small luxuries like tea and household goods and gave them some security in years of a poor harvest when grain was scarce and expensive.

> In Orkney every consideration is sacrificed to kelp. Agriculture is now very much generally neglected. Less grain is raised than was raised 30 years ago . . . The landowners of Orkney will find, too late, the great imprudence of thus neglecting the cultivation and improvement of their lands.
>
> Since the introduction of kelp manufacture, a great change has taken place in the state of society in Kirkwall. Country gentlemen have thus acquired, from their bleak estates, sums of money, great beyond all previous experience.
>
> *A Tour through Some of the Islands of Orkney and Shetland* (The naturalist Patrick Neill, 1776–1851)

The kelp season started immediately after the grain was sown and lasted from May to August. All available hands, mostly agricultural workers, were required to gather, process, and pack kelp for export so, to a large extent, crops were left unattended.

In the early 1800s mineral deposits discovered in Germany caused the demise of Orkney's industry, giving a wake-up call to the lairds left with money and land but little income. Neglected farms, hit by bad weather, produced poor harvests and little edible grain. The tables were turned, and it became essential for Orkney to import grain for food.

The situation is described in a letter from Robert Graham to his nephew in Edinburgh.

18th May, 1811
Watt of Breckness and Skaill
From Robert Graham

My dear Nephew,

The demand for beremeal is uncommon. I am certain most of them needs considerable supply. I mean the tenants as well as the other people through the neighbouring parishes.

Grain shortages, weather bad. Not been four days could be said to be middling weather the consequence been of serious concern to the people in general. Obliged to give grain for want of fodder to preserve their cattle from starving. Very great demand for oats and in particular bear meal which we have been grinding from the 'debt corn.'

Do you think you could purchase a cargo of bear so you could sell it when made into bear meal.

To William Watt Esqu
Andersons Lodging
14 Terrace,
Edinburgh.
(Kirkwall Library Archives, ref. D3/284)

On the land

The collapse of the kelp industry hastened the progress of land improvement in Orkney led by pioneer agricultural reformers such as Colonel Balfour, Alexander Graeme, Robert Scarth, Thomas Traill and others. The runrig system of separated strips of land was systematically replaced by the 'planking' system, where the land was squared into large fields. By the 1870s most Orkney farms were owned and run by estates or let to tenants, and, as a result, a farming community developed which was interested in modern methods of land management. One such was the application of a new fertiliser, Peruvian guano, imported via Liverpool.

Warren and Cumming, bakers, merchants and shipowners had premises in Kirkwall. The letter below was written by, or on behalf of, my great-great-grandfather Thomas Warren (letter from Warren and Cumming 17th January 1871).

To Mr James Johnstone, Orphir

Sir,
Enclosed you will find a/cs for the past year which please to examine and see if correct. You will observe we have charged no sacks. Peruvian

Guano was dear last season this season dearer still go on feeding. Prices for Feb will keep up this year on account of the [Franco-Prussian] war. We offer you 20/- for 40b [bushels] oats 1/- up or down according to weight and [? illegible] all your corn [bere].

Yours Warren and Cumming
(Kirkwall Library Archives, ref. D15/12/17)

'What we grow is what we eat and we put it to every good use under the sun.' – *Notes on Orkney History and Folklore* by Ernest Marwick, Kirkwall Library Archive (D31/1/2)

Bere, known colloquially on Orkney as 'corn', grew on the infield to be milled or malted for the household, and better quality oats from the outfield were saved at cutting time to be milled into oatmeal. The remaining oats were bruised for cattle feed, fed to poultry or saved as seed which was stored well away from mice and livestock. As agricultural improvement progressed, the acres growing oats increased and those of bere declined. This is clearly seen from records dated 1885 which list 33,457 acres of oats compared with 5,018 of bere. Each farm-worker was paid annually in 'bows' or 'bolls' of bere or oatmeal – a bow or boll weighed 10 stone. The quantity earned depended on the worker's status, thus a grieve or foreman received more than a ploughman.

Hand-stripped bere and oat straw, called 'gloy', was woven into Orkney 'stray'-backed chairs, and the woven straw baskets called cassies and cubbies.

Cassies were used to carry grain, fish and other items, while cubbies were used around the house to store domestic necessities like spoons, peats and food for the hens! Twisted straw ropes or 'simmons' were made into mats called 'flackies', used to fasten straw thatch onto roofs, secured the tops of cornstacks (known locally as 'skroos') and made panniers for horses. Straw served as insulation, as a draught-excluder and as bedding for people, cattle and horses alike – the less scratchy oat straw preferred! Straw-stuffed boots kept feet warm and, last but not least, there were handmade straw shoes, the 'stray buits'!

The planting

Ploughing began the farm year at Candlemas (2 February) or as soon as weather permitted. Horses, oxen or occasionally men, pulled wooden and metal ploughs, then followed with harrows and hoes to prepare the ground. Furrow by cold furrow, it was arduous work, and even women took a turn.

'Whose crops would be cut first, whose would have the best yield?' was always the question. Rival farmers tried various ploys. Seed corn was sometimes sprouted before sowing in early April to give it a 'start'. One year late snow covered Orkney and many were caught out. Germinating seeds wait for no man! 'Dae ye mind that bad year when there wis six weeks o' snow in March?'

Hand-sowing, walking the fields casting seeds from a basket in rhythm, is laborious work. In 1850 this job became easier with the introduction of one-man seed 'broadcasting' machines. In particular, the aptly named seed 'fiddle' appeared, which released a measure of seed as the sower walked over the field, pulling back and forth on a horizontal wooden bow. Whether sowing by hand or machine, women standing at the end of the rig replenished the seed to allow the sowing rhythm to continue. Horse, and subsequently tractor-drawn broadcasters were calibrated according to the seed being sown. Bere was set at 'five grains per horse's hoof', which is equivalent to three bushels an acre.

In 1920 Kirkwall merchants J. & W. Tait imported the first tractor, believed to be a 'Case', to Orkney. It was bought by the farm of Saverock, St Ola.

'Tractors and horses may achieve the same result but they behave differently!' The story is told of Birsay farmer Jimmy Adamson's first tractor. Out for a drive in his new green Fordson, he discovered that shouting 'Woa, min, Woa!' at the top of his voice did nothing to halt tractors after all. Fortunately he and the tractor survived the ensuing bump . . .

The hairst

All hands to the fields!

Bere usually ripened and was gathered in first. It can be difficult to cut because the crop 'lodges', which means it is easily broken by wind and rain as it ripens. The spiked seed-heads of the flattened crop burrow into the ground and begin to re-germinate. It is a strange sight to see green shoots appearing in the midst of a crop at harvest.

'Stooks' of six sheaves set up in the field to dry

Originally grain was cut by hand using a tool of pre-historic design, the 'heuk' or sickle. It was back-breaking work, bending to cut with one hand while holding stalks in the other. As J. T. Smith Leask recorded in his book *A Peculiar People*, 'Boy, boy, it waas coorse, coorse wark an sair, sair api' da back!'

Farmers, keen to speed operations, used ploys such as the 'mullewe' bere bannocks, which were dispensed to the reapers at the start of cutting. No sitting was allowed, each cast his bannock ahead of him, 'cut to it', took a bite, cast it further ahead and so on.

The long-handled scythe or sye, as it is called on Orkney, came as a welcome improvement to the heuk. A skilful scythesman could cut up to an acre of standing crop in a day, enough to employ two following workers who lifted and tied the sheaves. A sharp blade was essential to cut with ease. Scythes were sold in separate parts so it was easy to replace blunt blades from a back-up stock of sharp ones. Blades were frequently re-sharpened during the harvest. In good weather, cutters worked from dawn till moonlight, rewarded with a thick spread of 'heuk butter' on their bannocks.

Horse or tractor-drawn, the mechanical reaper harvested grain with cutting blades which ran at right angles to the wheels. A divider guided the crop into the cutting bar to feed the 'sheaf board' worked by a 'sheaf man'. When there was sufficient gathered to make a sheaf he dropped the board, pushing the pile to the ground with a rake as the board lifted back up to catch the next bundle.

Harvest was a time of camaraderie and banter. All those who were able took to the fields, competing in pairs to see who could lift, bind and stook sheaves to the row end first. Each farmer knew his crop by the band securing his sheaves, which was devised by twisting two handfuls of stalks into a long straw strap. Making this 'identity band' correctly was important because it was also essential that the sheaf stay intact till threshing. Six at a time, sheaves were built into stooks and left to dry in the field; then they were carted to the stackyard, piled into 'disses' (small stacks) and finally built into large stacks called 'skroos'. The heads of grain faced inwards and the straw outwards.

Farmers took pride building their cornstacks so that they not only looked good, but more importantly 'ran off the water' to protect the grain inside. Many won medals for their skill in building. For example, the illustrator Ruth Tait's grandfather, William George Rendall of Skaill Farm at Rendall, won several medals, one of which was presented to him in 1929 by the Orkney Agricultural Discussion Society.

There is a knack to building a skroo (cornstack). As the stack grew the man in the yard below was required to pitchfork sheaves more slowly to give the 'top builder' time to complete his unique design. However the man below did not always adjust his pace! 'When that man was forkan tae me at the top o' the stack he nivir slowed doon, he jist fired them ap as fast as he could, so I just fired them doon the ither side til I got the top feeneeshed.' (A comment by an Orkney farmer recorded in the Orkney Heritage Society Fereday Project, archive D70.)

A New Year song from the parish of Firth celebrates the cornstack building:

'Guid be tae this beurdly biggin
Frae the steeth an' tae the riggin!
We wis' mony stacks abune yer style,
Some for maut an' some for meal.'

Harvesting tradition and folklore

It was thought that the weather at Beltane on the first of May gave an indication of the harvest to follow.

> 'If the wind is sooth,
> There'll be bread for every moooth;
> If the wind is east,
> There'll be hunger for man an' baste;
> If the wind is west,
> The crop'll be lang an' slushy;
> If the wind is nort',
> The crop'll be short an' trig.'

A good harvest was vital, a poor harvest meant hardship.

Orcadian historian and writer Ernest Marwick recorded several years of poor harvest:

'So great is the famine that the people of low estate have nothing and those of greater rank nothing they can spare.' (1634.)

'At this time Orcadians fed on rumex weeds [docks and sorrels] and sea fowls, people died in hundreds, lands were laid waste for want of strength of man and beast.' (1680–84.)

'Many years of poor harvest and eleven of famine.' (1696–1783.)

'Very wet, little harvest.' (1815.)

'Year of the short corn, there was no rain from seedtime to harvest. Straw so short it had to be pulled up by the roots.' (1826–28.)

Notes on Orkney History and Folklore, 'Ill Years and Famine Years' by Ernest Marwick, Kirkwall Library Archive (D31/1/2).

Since Orkney was no stranger to famine, Orcadians made use of what they had and enjoyed it. They never burned straw or food on the fire and washed dishes in scalding hot water without soap so that food scraps were poured into the hen's tub with the water.

Various traditions were associated with the harvest. Some may appear contradictory, because each parish had their own ways of celebrating, but here are some examples.

The 'casting of the heuks' (sickles)

At the last cut of harvest the reapers tossed their heuks over their shoulders, chanting:

> 'Whar'll I in winter dwell,
> Whar'll I in vore dell,
> Whar'll I in summer fare,
> Whar'll I in harvest shaer.'

The way the heuk landed pointed the direction its owner might possibly next find work, but, if the point stuck in the ground it was said that that man would be 'in the kirkyard' before the year was out, a self-fulfilling prophecy perhaps.

The spirit of the corn

There is an ancient belief that the 'spirit of the corn' lives in the last sheaf and dies the moment it is cut. To avoid being responsible for this demise, more superstitious reapers cut the last sheaf by 'remote control', throwing their sickles, in a motion similar to skimming a stone across water, at the sheaf until the job was done. The sheaf was hand-bound and hung in the rafters till New Year, when, to keep the spirit alive through the next year, some was fed to stock and the rest ground into meal to bake the 'harvest bannock.'

The man whose sickle had cut the last of the grain and completed the harvest was given the best seat at the Harvest Home when he ate the bannock to save the crop spirit, and so redeemed the situation.

The straw 'bikko'

In some parishes, a straw bikko (dog) woven from the last sheaf straw was used to single out the person responsible for its cutting. Placed on the roof of his house, left at his door or tied to his cart, it was seen as a public insult. On Rousay the bikko was replaced by a straw man. More recently young women wove the last sheaf into an elaborate knot and tied wisps of newly cut corn into small knots to wear on their bonnets or give as a gift.

Feeding the birds

To feed the birds and small animals a corner of the last field was often left uncut.

The Muckle Supper

The farmer threw a 'Muckle Supper' to thank all those who helped with the harvest. A substantial meal, it may have been named after the Norse 'mikill' which means

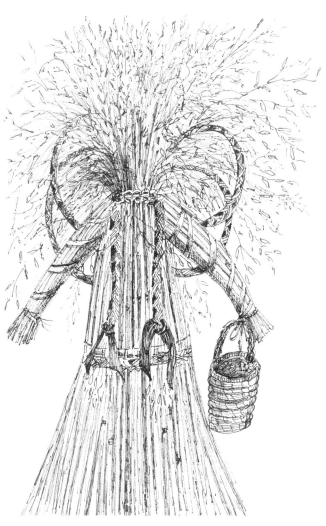

A straw dolly woven as part of a harvest celebration.
An ancient custom which began with the weaving of
the straw bikko

large. However, Ernest Marwick thought 'muckle' was a corruption of 'Mikkel' or 'Michaelmas', the feast of St Michael, which is on the 29th of September. John Firth in his *Reminiscences of an Orkney Parish* (1920) recalls these meals: 'When the last sheaf had been gathered in, the farmer had a muckle supper or harvest home. The pig or a sheep was killed, puddings were made, fowls roasted and ale brewed. Then his friends and neighbours were called in to join him in "rejoicing with the joy of harvest". But this high living was not of long duration, for the rent had to be met, and money was scarce.'

Mechanising agriculture

In 1898, Orkney's first binding reaper was demonstrated at Barnhouse Farm, Stenness.

The scene is described in *The Way Hid Wis* on Patricia Long's website *www.aboutorkney.com*: 'It was "like a market", older men picking up the sheaves, examining the knot, shaking their heads and muttering, "Wark o' the Devil".'

The safe anchorage of Scapa Flow played a strategic part in both World Wars when thousands of men were stationed on the islands. The population of the islands tripled during World War Two, creating a huge demand for local produce, and farmers, paid subsidies under the Government 'Plough-Up' scheme, reclaimed land to grow more crops and graze cattle. Under a regulation introduced by the Ministry of Food and Agriculture, millers were required to purchase grain from the farmers, mill it for them and sell it back. This, it seems, was purely a paper exercise. Miller Rae Phillips remembers filling in such paperwork as late as 1966!

'In 1959 Ted Zawadski brought the first combine harvester to Orkney, which revolutionised farming.' From *Willie's World 1959: The Diary of William Harvey* (2008), edited by Gregor Lamb.

Scarths imported both the Zawadski machine and one other bought by Louttit of Biggings at Rendall. On his retiral he sold the harvester to Ned Spence, who used it to start Orkney's first agricultural contracting business.

Threshing

> Now the golden grain
> Falls like April rain:
> Chaff, like floating mist
> By the sunshine kist,
> Brings again the dawn
> Of May mornings gone.
>
> Idly drifting by
> To the roof's dim span,
> Straw-clouds come and go:
> In the barn below,
> Ghostlike, moves a man,
> Raking down the sky.

From *Winter Threshing* by Robert Rendall
(1898–1967).

Bere is covered in a prickly beard called 'skegs', which made threshing an itchy, scratchy experience. Small amounts of bere for household use were threshed by lashing the heads against a hard surface to dislodge the grain. The main operation took place in the barn on a pre-scrubbed scrupulously clean floor where sheaves were laid head to head and pummelled with the swinging beaters of wooden-handled flails to detach grain from straw and chaff. Opened doors at each end of the barn created a current of air which blew the lighter chaff and dust away, while the heavier grain fell to the floor. Thankfully, by the 1850s, this dusty, back-breaking work was made obsolete by hand-operated and subsequently power-driven threshing mills. A travelling mill serviced small farms, while larger growers had their own equipment.

Grinding the grain: querns

A quern (Norse *kvern*) is a simple grinding mill consisting of two stones, the upper of which is rotated or rubbed on the lower one. Neolithic people ground grain on a 'saddle' quern using a stone 'rubber' held in both hands by the kneeling operator. This put pressure on their hips and knees with painful results. The gritty flour so painstakingly produced was certainly not 'tooth-friendly'.

The archaeology website *www.odysseyadventures. ca* records the discovery of one such quern: 'A large stone saddle quern and two grinding stones were found at the Knap of Howar on Papa Westray, one of the oldest preserved stone dwellings in North Europe and dated 3,600BC.'

The more user-friendly rotary quern was introduced during the Iron Age. It consisted of two circular stones, the lower fixed, while the top stone, rotated by a wooden handle, ground grain fed through a hole in its centre. This was collected as a coarse flour when it fell out from the sides of the stones. It is from this task that the saying 'daily grind', meaning labour that wears you down, has its origins.

Water and wind-powered mills come to Orkney

The milling of cereals was revolutionised by invention of the water-powered mill which was driven by vertical or horizontal wheels. It is uncertain when they arrived on Orkney, but, rental records of 1492 show mills were then well-established. By the seventeenth century there were at least 50 water or wind-powered vertical and horizontal wheeled working mills on Orkney.

In 1625 the Sheriff of Orkney issued a decree, at the request of William Sinclair of Sebay Mill, to confiscate querns and prevent avoidance of 'dues' because 'feudal suckeners' of the mill were 'home-grinding'. This and similar decrees had little effect and querns survived. 'Sucken' or 'thirlage' was the mill-owner's right to compel the tenants of his district, his 'suckeners', to bring all their grain to his mill. 'Mill-suckened' tenants were bound to this agreement, paying 'multures' in return to the mill-owner and were also required to participate in maintenance of the mill.

'Ye sookaners o' the Mill o' Bae
Come tae her the morn wi' simmons an' strae.'
(Traditional song quoted in Alexander Fenton's
The Northern Isles: Orkney and Shetland, 1978.)

Prior to the 1870s, grain was kiln-dried on the farm before being delivered to the mill, but this made it difficult for the miller to control moisture levels. To rectify this, kilns were built into new or added onto existing meal mills. These kilns are easily recognised because of the rise in the mill roof at one end.

The story of Barony Mill

The Barony Mills are built on the farmland of north-west Birsay and take in three water-powered meal mills and a separate threshing mill.

As the seat of power during Norse and medieval times, this area played an important part in the history of Orkney. The Viking earls resident in Birsay Palace sourced food from the surrounding farms of their domain which they called 'bordlands'. They were succeeded by the Stewart earls who, in order to supply their households, took possession of the 'bordland' farms to create a single large 'estate' farm and named it 'Boardhouse'.

The loch supplying water for the mill dam was first called the Loch of Twatt, then Loch of Kirbister and finally Loch of Boardhouse. It is fed by a burn, said to be the longest on Orkney, which flows from Harray where it is called Burn o' Rusht. The burn's name soon changes along its course into Bog o' Surtan, then Burn

The water wheel at Barony Mill

o' Kithuntlins, Burn o' Grid, Burn o' Lushan, Grip o' the Sty, Burn o' Kirkgeo, Burn o' Gyron, Burn o' Teeoma and Burn o' Hillside; it enters Hundland Loch as Burn o' Durkadale and emerges as Burn o' Kirbister, then runs into the Boardhouse Loch which feeds the mill dam. Its work done, the burn leaves the loch as Burn o' Hunto, to end its journey in the sea at Birsay Bay. Phew – some mill race!

Set on a corner of the Barony site is the original 'click' mill which was built around 1500. A click mill is operated by a vertical paddle, through which water is channelled into the side driving the wheel to turn the upper millstone. The 'click' is caused by a wooden

An old drying kiln

peg inserted into the top outside edge of the millstone which hits a protruding part of the cradle or shoe, a hopper that feeds small amounts of grain into the rotating stones.

A smaller mill on the site had no kiln. Subsequently, a threshing mill was built to service the nearby Boardhouse farm. Although it was powered from the same mill-lade it had no direct connection with the mills.

An Edinburgh lawyer, Lord Dundas, who owned the ground, was responsible for building the third mill. This was designed as a taller building, incorporating a kiln. Building started in 1872 and, with the help of the community, farmers and stonemasons, the Barony Mill was completed in one year. The first miller, William Hepburn, began milling grain in 1873 and the mill continues to operate today.

Because it was the largest building in the area it was put to other uses, being much in demand for weddings and dances. Homebrew and many a celebratory Orkney cog fuelled such exuberant dancing that the walls began to bulge outwards. To rectify the problem metal rods and plates were fixed into the walls and the revelry continued. (An Orkney cog is a round hand-crafted wooden drinking vessel with handles which is filled with a potent punch-type brew and shared at gatherings like a communal cup.)

Millstones

There was a tradition whereby millers were required to provide their own stones. However most stones would outlive the miller. To a certain extent this continues today because there are few millstone quarries, and the miller alone has the knowledge to select stones suited to grind a particular grain. When the Barony Mill was built in 1872, a miller could only earn a maximum of £50 in a good year. Considering that imported new French burr stones cost £33 each, a rethink was needed. In 1873 secondhand French burr stones were sourced from the mainland and remain in use today. Cheaper stones were bought from Clackmannan for £24, and others, suitable to mill bere, were bought locally at £12 a unit. These stones came from Quoy Ayre, Yesnaby,

A French burr stone, made from chalcedonic hornstone quartz. A pair of such stones bought in 1873 are still in use in the Barony Mill today

the only millstone quarry on Orkney, which supplied the islands' mills for centuries. The last stone, made by Jock Marwick and his father, was shipped to Trenabie Mill, Westray in 1927. At that time millstones were charged at £1 per inch thickness at the 'eye' (the hole in the centre of the stone) plus 10 shillings yearly rent per millstone, payable to the Breckness estate.

The uneven, often boggy terrain of Orkney was only suitable for two-wheeled horse-drawn vehicles, therefore it was impossible to transport a millstone by that means. A stone thus loaded would lift a harnessed horse clean off the ground! The solution was to turn it into a wheel. A wooden plug secured the 'mill-wand,' a 9ft to 10ft-long wooden pole, through the eye of the stone, and correspondingly wide gaps were left in boundary dykes to facilitate the stone's journey. Three strong men holding each side of the wand rolled the stone over flat ground but, to pull the stone uphill, a horse was harnessed to a U-shaped iron loop attached to the centre of the mill-wand while the men at each end of the mill-wand kept the stone upright. At the top of the hill, the harness was reversed and the horse walked behind the stone to hold it back from rolling out of control. The men held the ends of the wand to keep the stone vertical, but were ready to 'coup' (tip) it on its side into a field or crop should it start to run away. A millstone weighs around one ton, so a runaway could do a lot of damage. A rolling millstone gathers no moss, but could flatten a house!

An Orkney riddle
'Twa grey galts [pigs] lying in a stye
The mair they get the mair they cry.'

Solution: a pair of millstones.

Restoration of the Barony Mill

Modern equipment and farming practices encouraged farmers to increase the acreage of two-row malting barley they grew to sell for whisky and beer production. By 1982 cultivation of bere had decreased dramatically, prompting retiring miller Olive Flett to write:

'The Boardhouse Mill must soon close doon
If hid's no fixed pretty soon.
Whit a job hid wad be,
If there's no bannocks for the tea.
So grow your corn and a'll mak meal
And a'll guarantee ye plenty of sale.
Some might think its awfa dear,
But it'll do more good than a bottle o' beer.'

The Fletts sold the mill to Fergus Morrison in 1982. He continued to operate the mill after he sold it to Orkney Islands Council in 1986; renting the premises to continue milling until he moved south to Golspie in 1992.

The mill then lay vacant, prompting three Orcadians, the late Sandy Scarth of Twatt Farm, Keith Johnson and Johnny Johnston to make a plan to preserve both the mill and Orkney bere. The Birsay Heritage Trust was formed in 1998 to restore the running of the Barony Mill and officially took over that same year. To help increase the acres under bere cultivation, they sourced one tonne of grain from a number of local farms. A quarter tonne was retained for seed, the rest milled. By 2004 there was little or no cultivation of bere in mainland Scotland, but on Orkney bere was making a comeback. Although it is unlikely to be grown again on a scale seen in the eighteenth and nineteenth centuries, increased interest in this wholesome ancient grain, plus improved growing practices, will enable this important crop to be grown and processed commercially. An increasing acreage of

The hopper feeding grain down into the oatmeal stones encased below

bere is grown on a number of farms by an agricultural contractor employed by the Birsay Heritage Trust. The Agronomy Institute, Orkney College UHI, grow bere on their Weyland farm in Kirkwall. They also employ farmers across the island to grow the crop for them.

Milling bere

It takes 2½ days to dry, cool and grind one tonne of bere into beremeal. It is done in four stages.

Grain at 14% moisture is further kiln-dried to a moisture content of 9–10%.

Twelve hours later the cooled grain is passed through a set of old millstones which have been coated with corundum, known as 'shelling stones'. They are set just lower than the size of the grain in order to crack the outer husk and free the kernel. The coom (dust) is removed by an eight-feet-long sieve. The mix of kernels and husks then goes through two 'fanners' which blow away the husks, known as 'scrubs', which are used to feed the kiln fire.

The cleaned grain is then passed through a pair of French burr-stones and ground to a coarse meal known as 'grap'.

For the final stage of milling, the original local stones from the former quarry at Yesnaby have been replaced by stones from Derbyshire. The height of the stones is subsequently adjusted to produce fine, medium or coarse meal. The meal is passed through sieves to remove the 'sids' (the inner skins) and bagged. The sids also feed the kiln fire.

Sharpening and dressing the grooves in the stones is still done by hand using a mill-pick, a form of chisel with a heavy wooden handle, and an electric diamond disc cutter. Afterwards resetting and adjusting the millstones is a time-consuming, highly skilled job carried out by the miller and apprentices.

What of the mill today? Still in good heart, run totally by water, the mill jumps to life the moment the sluice is opened. It has changed little since 1873. The original 'tilley' lamps were replaced in 1956 by electric

Bagging the meal

light powered by a small generator, and in the 1980s changed to mains electricity. A toilet was installed at the same time. In 2009 the building was re-roofed and re-pointed by the council. Grain is milled during the winter when there is plenty of water to fill the mill race to power the wheel. In spring and summer the mill is open to the public who can see the miller demonstrate the mill machinery at work.

The Future: The Birsay Heritage Trust

The Trust raises awareness of beremeal by advertising in tourist publications and attending events and trade shows. The Speciality Food Show 2014 at Glasgow SECC was a steep learning curve, but a successful one. Bere is now listed with several wholesale distributors, including Green City, Real Foods of Edinburgh and Highland Wholefoods. The Barony Mill won best product runner up award for bere bannocks baked by miller's wife, Margaret Phillips.

Inside the Barony Mill store

Attractive new packaging, different pack sizes, a small guide and recipe book have raised the grain's profile, along with a listing in Slow Food Scotland's 'Ark of Taste', which aims to save endangered traditional foods. The Trust employed Leatherhead Research to carry out tests to identify the characteristics of beremeal. Edinburgh's Queen Margaret University carried out nutritional analysis and developed a sourdough bread on behalf of the Trust. Stockans of Stromness also created an award-winning beremeal oatcake in collaboration with the university.

Now bakers on and off Orkney are incorporating beremeal into nutritious new breads and biscuits. Brewers on Orkney and beyond are brewing bere ales and distillers bere whisky, gin and vinegar. Restaurants and cafes are serving beremeal breads, bannocks and baking, and discerning cooks and home-bakers continue to discover the goodness of beremeal in the kitchen.

Scrub (the outer layers or husk) left from milling bere is also used in the manufacture of local pottery and soaps.

Research into bere

Research into bere continues in the Agronomy Institute at Orkney College, which is a partner in the Northern Periphery and Arctic Programme Cereals Project. DNA and genetic analysis of bere is being carried out by the University of Manchester, the James Hutton Institute, the Science and Advice for Scottish Agriculture division of the Scottish government and the John

Innes Institute. Several nutritional analyses have been completed, the most recent by the Rowett Institute of Nutrition and Health, University of Aberdeen.

Results of new research were reported at a conference on Orkney in June 2017. It is predicted that morphometric and DNA-based approaches will help identify grains of bere. Initial results suggest that bere has been present on Orkney since Neolithic times.

Bere harbours genetic diversity with traits that potentially will help crop development to tackle issues such as climate change and food security. Tim George of the James Hutton Institute commented: 'It seems that bere has been saved from the brink of extinction and those precious grains left by our ancestors all those years ago may give us some tangible benefits today in supporting rural communities, human health and environmental and agricultural sustainability.'

Why eat Orkney beremeal?

Orkney bere is grown from seed directly traceable to local growers of the 1970s and has not been imported. With the added bonus of growing in Orkney's fertile soil and pollution-free sea air, it has a unique provenance.

Nutritional analysis of beremeal has revealed:

- It contains a wide range of macro and micro nutrients, particularly folate (vitamin B12), thiamine (vitamin B1), pantothenic acid (vitamin B5), iron, iodine, zinc and magnesium.

- It is high in beta-glucans (soluble fibre), which

is valuable in the diet as protection against cardiovascular disease, colorectal cancers, type 2 diabetes and in weight management.

- Recent results comparing bere with modern barley varieties show that bere, although lower in beta-glucans, is higher in fat, sugar, protein and minerals, notably magnesium, phosphorus, calcium, manganese, iron, copper and zinc.

- Beremeal contains higher levels of iodine, magnesium, iron and B vitamins compared with white wheat flour.

- Research by the Rowatt Institute of Nutrition and Health concludes that bere contains potentially beneficial health properties. It has unique metabolic, beta-glucan and soluble fibre profiles that could give consumers substantial health benefits, including anti-inflammatory properties and improvements in cardiovascular function.

Orkney's unique ancient grain, is, without doubt, nutritionally valuable. It is a wholesome 'time capsule' left by our ancestors to nourish the future.

Time-honoured preparations of bere: brewing and distilling

'Stromness ought to be honouring William Clark and his good wife Mareon more than anyone. If they weren't the first Stromnessians, they were the first recorded business people in the parish. With forethought and some daring, William saw the ships of Europe growing larger with every decade and venturing further north and west, stormbound often between Brinkies and the Holms. William, with the Earl of Birsay's permission, built an inn at the very tip of the harbour. You may be sure, on a stormy winter night, half the languages of Europe were spoken across William's barrel of ale and Mareon's bannocks and cheese!'

From *Under Brinkie's Brae* by George Mackay Brown (1979)

The inn was built in 1595.

Traditionally made from kiln-dried malted bere, ale was generally brewed by the women of the house. On Orkney there is a long tradition of excellent home-brew, each one unique to the brewer.

A young wife home alone was caught unawares by a visit from two excisemen with no time to hide her illicit malt. It was unheard of to allow such men over the doorstep, so she took them by surprise inviting them in for a glass of ale. They accepted so she made them dinner. Neither party mentioned the malt. 'She was lucky that day.' (The Orkney Heritage Society Fereday Project, archive D70.)

Historically, early whiskies were distilled from malted bere but this fell out of use because it produced a lower alcohol yield compared with modern barley cultivars.

Hugh Yorston 17th April 1797.
To William Watt esqu or Breckness and Skaill

I wrote you saterday last week requesting some malt for whiskie to yourself – if you will please to give me bear I shall send it for malt and distill it as quickly as possibly for you and if you will please to let me have the quantity of bear I wrote you of or so much as you can conveniently spare I shall either give you whiskie therefore or shall at Lammas or Martinmas pay your current price for the quantity you can let me have and I would wish to have as soon as possible.

I am greatly obliged to you for the potatoes and shall pay you the price soon.

With best wishes, I am respectfully,

Sir, Your most obedient servant,

Fobbister April 1797 Hugh Yorston

(Kirkwall Library Archives, ref. D3/43.)

In 1798 Magnus Eunson founded Highland Park, Orkney's first distillery, so named because it is built at 'High Park' on the outskirts of Kirkwall. Here 'whiskie' was distilled from bere barley malted on site, using peat cut from Hobbister Moor to dry the malt.

'Bout vines an' wines, an drunken Bacchus
An' crabbit names an' stories wrack us
An' grate our lug
I sing the juice Scotch bear can mak us
In glass or jug.
    From *Scotch Drink*, Robert Burns

The old Orkney kitchen

A typical old Orkney kitchen is described in John Firth's *Reminiscences of an Orkney Parish* (1920).

> Cooking was of the simplest kind. The only utensils in use were the brand iron for bannock baking, the yetlin of cast iron for baking scones and three-toed pots of different sizes, varying from the muckle pot of ten gallons to the peerie pot for the bairns' gruel. These cooking utensils were suspended over the fire by a long chain or by four folds of straw simmons wound together, with five or six iron links next to the fire. The pots were hung from this suspended by a crook, which was linked up or down according to the degree of heat required for cooking. The lum was the only outlet for smoke which "eddied" and whirled through the air of the room – it may be that it acted as a disinfectant as well as a deodorizer where man and beast were herded together in such limited space!

A low wall separated livestock from the 'living space'. This meant there was no peace – imagine living side by side with hens, calves, ducks, geese and the odd pig! The 'ben end' or 'master bedroom' was also used to store potatoes, a meal girnel, malt, a quern and a small closet, the 'ale hurry', which held stone pigs (jars) of ale.

A meal girnel is a deep strong wooden chest into which meal was tightly packed to keep over the winter. Girnels were often divided in two, one part for beremeal, the other for oats. In the meal girnel at Corrigall Farm Museum a chequer board is cleverly inserted halfway down the central partition. No game was possible until both compartments were consumed to a suitable level.

BERE RECIPES

Kale and knockit corn

The 'knocking-stane' was used to de-husk bere. It was a large stone block with a cup-shaped hollow in the centre, into which the wholegrain, softened in warm water, was poured and then knocked with a mallet or stone. The husks were poured off with the water leaving 'knockit corn', an Orcadian pearl barley.

Fresh knockit corn was mixed with boiled cabbage or kale and eaten with relish.

> Thoo's be kissed, an I's be kissed,
> An' I's be kissed the morn,
> But the sweetest kiss that e'er I got
> Was kale an' knocked-corn.

From John Firth's *Reminiscences of an Orkney Parish* (1920).

Broths and soups with knockit corn

Knockit corn thickened soups and broths. A straw band was laid round the pot's rim to trap floating husks and was whisked across the surface to skim off the remainder before dishing up – this was known as 'pufflan' the pot.

Burstin

Burstin is a meal made from toasted bere. The 'mettins' (grains) were 'stir toasted' in a muckle pot at the fireside, then quern-ground, sifted and stored for summer use.

Burstin mixed with sour milk (louts from the Norse *lut*) made a curd-like dish which was eaten as a 'cooling sequel' to dinner.

Burstin brose was made by stirring burstin into hot water with pepper, salt and a lump of butter and supped immediately.

Hints and tips on cooking with beremeal

There is health and flavour in every handful.

- Seasoning: the earthy flavours of beremeal reduce the amount of salt required to season dishes. It has a robust taste, so be cautious till you find what suits you. Keep it simple when adding herbs and spices.

- Thicken mince and gravies and toss meats in seasoned beremeal for casseroles and stews.

- Coat meat, fish or vegetables, adding herbs or spices if liked. Grill or fry till crisp.

- To make fish fingers, cut a meaty fish like cod, tusk or pollock into fingers or strips, toss in beremeal seasoned with ground black pepper and shallow fry.

- For use in a stir-fry, toss strips of beef, pork or chicken in seasoned beremeal before stir-frying.

Baking with beremeal

Beremeal is low gluten, not gluten-free. It is easier to handle if mixed with a proportion of flours with more gluten in them.

Add beremeal up to one third of total flour weight to make bread in a breadmaker or by hand. It will have a closer texture than other breads. Try adding a handful of seeds, rough grains or oatmeal to open the structure.

Bere bannock makers generally use equal quantities of bicarbonate of soda and cream of tartar. The use of buttermilk or sour milk produces a light texture (fresh milk is easily soured with a spoon of fresh lemon juice). Use a girdle or thick-bottomed frying pan. Bake on a low to medium heat to let the bannock bake through and avoid undercooked middles!

To use in general baking, replace 1 tbsp flour with 1 tbsp beremeal and ½ tsp baking powder to keep a light texture. Beremeal tends to be dry, so aim to keep moisture in soft baked goods by adding other ingredients. Soft brown sugar, syrup and honey, more liquid, mashed banana or dates can be added to help keep a moist texture.

Bananas, chocolate and walnuts marry well with beremeal.

To use in making pastry, replace one third of flour weight with beremeal.

Bannocks

'The girdle until recent times took the place of the oven, the bannock, the loaf.'

F. Marian McNeill, *The Scots Kitchen* (1929).

Originally bannocks were unleavened round flat cakes of barley or oatmeal, baked on a flat stone that was heated in the fire; this was later replaced by the girdle. Known on Orkney as a 'yetlin' or 'yetleen', a girdle is the round flat metal plate used for baking, which once hung on a 'cruik', a hook over the fire. Latterly bakers have added flour and raising agents to bannocks to produce a lighter texture.

Left-over bannocks were crumbled and mixed to thick porridge with warm milk; or were split, buttered and toasted over the fire.

Early bere bannocks

Catherine Brown, Scottish food writer and food historian, has kindly given me permission to include these historical bere bannock recipes from the early 1800s which pre-date the use of chemical raising agents. They are taken from her *Scottish Cookery* (Birlinn 2014).

Beremeal bannocks (unleavened)

These are somewhat akin to an Indian chapatti since they do not rise. The charm lies in the barley flavour and in the contrasts of texture; the soft inside and harder, but not quite crisp outside.

Makes 1 bannock 28cm (11in) diameter or 2 small bannocks the size of a hand, which are easier to handle.

300ml (½ pt) milk
25g (1 scant oz) butter
115g (4oz) beremeal
¼ tsp salt

Put the milk into a pan, add the butter and heat to melt, add salt. Stir in the meal; it should come together into a soft paste. Turn out onto a beremeal-dusted board and knead lightly to a smooth ball, dusting well on top with beremeal. Roll out with a rolling pin into a round about 28cm (11in) in diameter and about 5mm (¼ in) thick.

Heat a girdle or thick-bottomed frying pan on medium heat. Test with a dusting of beremeal: if it burns the temperature is too high. Grease with oil. Slide a rolling pin under the bannock and lift onto the girdle. Bake about 5 minutes till browned on one side, then turn using a large fish-slice, or cut in two and turn both halves separately. A wooden bannock 'spathe', about twice the size of a fish-slice, was used in the past for this. Smaller bannocks turn more easily. Bake the other side for 5 minutes till browned, cool on a wire rack and eat warm with butter and cheese.

Beremeal bannocks (unleavened, wafer-thin) ⚲ ⚲ ⚲

Another early form of bannock was made with a much thinner mixture poured onto the girdle like large

Breton pancakes (crêpes). Franco-Scottish connections may have had something to do with this version, since there is a striking resemblance between the large Scottish girdles and the large iron plates used in Brittany to make crêpes. In his *Travels in England and Scotland in 1799,* Faujas de St Ford talks about 'barley cakes, folded over', which probably referred to these bannocks.

Makes 6 to 8 small or 3 to 4 large bannocks

115g (4oz) beremeal
3 eggs
25g (1 scant oz) melted butter
Water to mix

Heat the girdle or thick-bottomed frying pan on a medium heat till hot. Sift the beremeal into a bowl. Beat in the eggs, melted butter and enough water to make a thin pouring consistency. Put into a large jug with a good pouring spout. Oil the hot girdle. Pour batter onto the girdle and spread with a spatula to make a very thin round pancake. Cook till browned on one side, turn and cook on the other. Cool in a tea towel on a wire tray. If using a frying-pan, pour in just sufficient batter to coat the base of the pan by tilting in a circular motion.

Serve spread with butter, crowdie mixed with cream, cream cheese, honey or preserves. Roll up into a stick of rock shape or fold into four layers. Enjoy freshly baked and warm. They can be frozen, wrapped separately and sealed in a polythene bag, for up to two months.

Auntie Betty's bere bannocks

From Val Johnston of Ruebreck, Birsay. 'I make bere bannocks most weeks, the recipe I use is one passed on to me by my late Auntie Elizabeth (Betty) Miller from Kirkatoft in Evie. Auntie Betty was my father's sister. She married the farmer Davy Miller and moved to Nigley Farm, Evie. When Auntie Betty retired to Kirkatoft nearby, she had more time to show me how to make bere bannocks, oatcakes and thin pancakes. I was newly married and home-baking saved precious pennies. Auntie Betty's family are delighted to share her recipe. She was a quiet, reserved lady but I know she would be pleased. In my mind's eye I can picture her secret smile of pleasure.'

Makes 2 bannocks

115g (4oz) self-raising flour
115g (4oz) beremeal
A good pinch of salt
1 tsp bicarbonate of soda
1 tsp cream of tartar
Water to mix

Sift the dry ingredients into a large bowl and mix together. Add enough cold water to mix to a soft smooth dough. Divide the mixture in two. Turn one half out on to a surface well dusted with beremeal. Gently shape into a bannock, handling as little as possible, and transfer onto a medium to hot girdle. The girdle should feel hot when you pass the palm of your hand over it, but not too hot or the bannock will burn. Experience is a good teacher. Cook for around 3 to 4 minutes and turn. Cook a further 3 to 4 minutes or until both sides are lightly brown. To test if it is cooked in the middle, tap the bannock, and if it sounds hollow, it should be cooked. Cool wrapped in a tea towel on a wire cooling tray. Repeat with the rest of the dough.

Val's tips

Make the dough as wet as possible. If it is too difficult to handle, shape the bannock on a sheet of foil dusted with beremeal. Use the foil to lift the bannock over the girdle and carefully slide it onto the hot surface. Don't worry if the bannock cracks, it will be moist inside.

Marion Whitelaw's bere bannock

A recipe from Ruth Rosie. Marion Whitelaw passed her bannock recipe on to her daughter Ruth thus: 'A cuttie bit o' butter, a kneve [handful] o' flour, 2 kneve o' beremeal, a grain of cream of tartar and a grain of soda. Water to mix.'

Ruth makes very good bannocks – 'they don't look neat and round, they have taste!' she said.

She worked on the Orkney ferry for over 20 years, then ran a popular catering van named Tea and Tabnabs. At sea a 'tabnab' is a break for 'a fine piece and cuppa'. Now retired, she still bakes bere bannocks and prefers to eat them a day old with a fresh boiled duck egg or partan (crab meat), simply seasoned with salt and pepper.

Makes 1 bannock

30g (1oz) plain flour
60g (2oz) beremeal
¼ tsp cream of tartar
¼ tsp bicarbonate of soda
½ tsp cooking oil
Water + ½ tsp vinegar

Heat a girdle on a low to medium heat. Sift the dry ingredients into a bowl. Mix the oil and vinegar with a little cold water. Mix into the dry ingredients with more water to make a soft dough. Oil the girdle. Shape the dough into a round with floured hands. Lay on the girdle and flatten with the palm of your hand. Bake for 4 minutes, turn and bake for a further 4 minutes. Cool in a clean tea towel on a wire rack.

Barbara Hewison's Westray bere bunnos

Barbara was born at Balaclava, a small farm on Westray, and learned to bake there, taught by her grandmother. She remembers making 'bere bunnos' with buttermilk so fresh it still had small pieces of butter floating in it. 'Bunno' is the name for a bannock on the north isle of Westray.

Barbara bakes bunnos, often twenty at a time, for the local Pierowall Hotel and, over the years, has baked many more for weddings, special occasions, family and friends. Full of energy at 80 years young, she still works with husband Marcus on their Westray farm and bakes on most days. She believes their health and energy stem from a lifetime of eating a simple diet. 'Must be all the bere bunnos we eat,' she says!

Makes 2 bannocks 15cm (6in) in diameter

115g (4oz) beremeal
115g (4oz) self-raising flour
1 tsp bicarbonate of soda
1 tsp cream of tartar
1 tbsp corn oil
Cold water to mix

Heat the girdle or thick-bottomed frying pan on a low to medium heat. Do not have it too hot. Sift the dry ingredients into a bowl and mix to a soft stiff dough with cold water. Turn onto a floured board and knead lightly. Cut in half. Roll each to a round 5mm (¼ in) thick. Bake on the girdle for 3 to 4 minutes on each side till browned and cooked through. Test by tapping the top with the knuckles; a hollow sound means it is baked. Cool wrapped in a clean tea towel on a wire cooling tray.

Davy Bain's oven-baked bere bread

Davy was born at St Ola near Kirkwall. When he was young, his father, who had served his time at a local bakery, taught him to bake. During his busy working life with the Milk Marketing Board, then as manager of a social club, Davy had little time for baking. Since retiring twenty-three years ago he has found baking success. His bere bread is a must on the menu at the

Peedie Kirk lunches that are served daily during the annual Orkney International Science Festival.

Makes a tray 30cm (12in) x 23cm (9in) and 5cm (2in) deep

290g (10oz) beremeal
140g (5oz) plain flour
1 level tsp salt
1 tsp cream of tartar
1 tsp bicarbonate of soda
25g (scant 1 oz) soft butter
200ml (7fl oz) plain yoghurt
400ml (14fl oz) soya milk (or use cow's milk instead)

Heat the oven at 220ºC/fan 200ºC/425ºF/gas 7. Oil and line a baking tray. Sift the dry ingredients into a bowl. Rub in the butter. Mix the yoghurt and milk together, pour into the dry ingredients and mix to a soft dough. Spread evenly over the prepared baking tray and bake for 10 to 12 minutes till golden, firm and springy to touch. Cool in the tin and enjoy fresh. This will freeze for up to 3 months.

'Bere-izza'

Created by the ladies of the Peedie Kirk, the 'bere-izza' has become a favourite.

Makes enough to serve 12

1 baked bere bread, split in half
1 jar 200g (7oz) tomato chutney

1 x 500g (1lb 2oz) passata (sieved tomato)
Salt and ground black pepper
500g (1lb 2oz) Orkney cheddar cheese, grated or
thinly sliced
6 firm tomatoes, sliced

Heat the oven to 180ºC/fan 160ºC/350ºF/gas 4. Mix
the chutney with the passata and season to taste.
Spread the cut side of the bread with a thin layer of
tomato mixture, top with cheese and sliced tomato.
Bake for 15 minutes till bubbling, cut and serve hot.

Any topping left over can be kept sealed in the
fridge for up to 1 week or frozen for up to one month.

The Postie's beremeal bannocks

From retired postman Johnny Johnston, who suggests
you should 'consume with copious amounts of ale plus
plenty of Orkney butter and cheese'.

Makes 2 or 3 bannocks

225g (8oz) beremeal
115g (4oz) plain flour
1 tsp bicarbonate of soda
1 tsp cream of tartar
Pinch of salt
Milk, water or buttermilk to mix

Heat a girdle on a low to medium heat. Sift the dry
ingredients into a bowl and mix to a stiff dough with
water, milk or buttermilk. Turn out onto a board dusted
with beremeal and flour. Half and shape each half into

a round flat bannock approximately 1cm (½in) thick. Lift the bannock onto the girdle. Bake for 4 to 5 minutes on each side. When the bannock sounds hollow when knocked gently with the knuckles it is cooked through. Cool in a clean tea towel on a wire tray. Enjoy fresh and warm or cool and freeze up to 1 month.

Alan Bichan's bere bannocks

Alan Bichan has written extensively about Orkney's rich larder, including his monthly article in the *Orcadian*. Regarding bannocks, he suggests 'Some bakers use twice as much beremeal as flour, but I find equal quantities of each a balance which appeals to many palates.'

Makes 1 bannock 15cm (6in) in diameter

50g (1¾oz) beremeal
50g (1¾oz) self-raising flour
1 level tsp bicarbonate of soda
1 level tsp cream of tartar
Pinch of salt
1 tsp cooking oil
Approximately 60m (2½ fl oz) water
Extra flour for shaping and cooking

Sieve all the dry ingredients into a bowl. Stir in the oil and add enough water to form a dough which leaves the bowl clean. Beat well with a wooden spoon, then shape on a floured surface into a 15cm (6in) round. Heat a girdle or thick-bottomed frying pan – the temperature should be somewhere between cool and medium hot. Dust with flour and cook the bannock for a few minutes till it is risen, turn over and bake the other side. If you want a perfectly soft bannock, wrap it in kitchen towel and a clean tea towel so that the steam does not escape during the cooling process.

Bere bannocks from the Creel Restaurant

For many years Alan Craigie was chef patron of the famous Creel Restaurant at St Margaret's Hope. This recipe is based on the bere bannock he served there.

100g (3½oz) beremeal
100g (3½oz) plain white flour
1 level tsp cream of tartar
2 level tsp bicarbonate of soda
2 tbsp natural yoghurt
Milk to mix

Heat a thick-bottomed frying pan over medium heat. Sift the dry ingredients into a bowl, then mix in the yoghurt with enough milk to make a very soft dough. Turn onto a beremeal-dusted board and cut in two. Test the heat of the pan with a dusting of flour: if it turns golden brown the temperature is correct. Lift one piece of dough into the dry pan, dust with beremeal and flatten to a round 2.5cm (1in) thick. Bake for 5 minutes on each side. Cool on a wire rack and serve warm.

As an option beer can be used instead of milk and yoghurt.

The Birdy Man's bere bannocks

During seven years as RSPB warden on Egilsay and Rousay, Andy Mitchell devised his own version of a bere bannock recipe that had been given to him by Rousay Pier Restaurant owner, Itha Flaws.

Makes 4 bannocks

175g (6oz) beremeal
175g (6oz) plain flour
1 level tsp salt
1 tsp cream of tartar
A drizzle (about 1 dessertspoon) olive oil
1 tsp bicarbonate of soda
120ml (4fl oz) natural yoghurt
120ml (4fl oz) milk

Heat a girdle or thick-bottomed frying pan on low to medium heat. Sift the dry ingredients into a bowl. Add a drizzle of olive oil. Mix the bicarbonate of soda, yoghurt and milk together till they are fizzing and mix with the dry ingredients to a soft dough. Turn onto a floured board and divide into four. Flatten each with a floured hand to a round shape approximately 1cm (½ in) thick. Bake on low heat for 5 minutes on each side until lightly browned. Cool in a tea towel on a wire rack and enjoy fresh.

Andy bakes two bannocks in this way, then further divides the dough into four smaller bannocks. He enjoys these warm from the girdle, halved and topped with a softly poached egg.

These bannocks freeze for up to three months.

Margaret Phillips' bere bannocks

Margaret likes to bake her bannocks with very little wheat flour and mixes them with whatever she has to hand – once, even, with a bottle of home-brewed ale!

Makes 3 bannocks 20cm (8in) in diameter

350g (12oz) beremeal
115g (4oz) plain flour
Pinch of salt
30g (1oz) margarine
1 tbsp natural yoghurt
Milk and/or water, sour milk or ale to mix

Heat a girdle or thick-bottomed frying pan on a low to medium heat. Sift the dry ingredients into a bowl, rub in the margarine and mix to a soft dough with yoghurt and milk or other liquid. Turn onto a floured board and divide into three. Dust each with flour, shape and put onto the hot girdle to bake 4 to 5 minutes on each side. Cool in a clean tea towel on a wire rack. Serve warm. These can be frozen for up to two months. Store fresh, sealed in an airtight bag, for up to three days.

Margaret uses an electric crêpe maker to bake her bannocks. This is quick and it is easy to adjust the heat.

Lily Russell's bere bannock

Orcadian friend Margaret kindly sent me this recipe from her Dad's sister, Lily, who lived at Cannigal farm, St Ola.

Makes 1 bannock approx. 15 to 20cm (6 to 8in) in diameter

30g (1oz) beremeal
30g (1oz) self-raising flour
¾ tsp bicarbonate of soda
¾ tsp cream of tartar
Pinch of salt
Water to mix

Sift the dry ingredients into a bowl and mix with enough water to make a smooth clean dough. Turn onto a floured board, dust the top with flour and pat out with the palm of your hand to a round 1cm (½in) thick. Heat a girdle or thick-bottomed frying pan on low to medium heat. Dust the bannock with flour and bake for about 3 or 4 minutes on one side till lightly browned, turn over and bake the other side. Cool in a clean tea towel and enjoy warm and freshly baked.

Maureen Flett's microwave bere bannock

This recipe is from my Orkney friend Margaret's cousin Maureen, Lily Russell's daughter. A recipe with potential.

Makes 1 bannock about 15cm (6in) in diameter

60g (2oz) beremeal
60g (2oz) plain flour
1 tsp bicarbonate of soda
1 tsp cream of tartar
Pinch of salt

Drizzle of oil
Milk or water to mix

Sift the dry ingredients into a bowl. Add the oil and mix to a soft dough with water or milk. Turn onto a sheet of beremeal-dusted kitchen towel on a flat, microwave-friendly plate. Dust with beremeal and flatten with the palm of your hand into a round about 1cm (½ in) deep. Cover with kitchen towel. Bake on full power for 2 minutes. Cool on a wire rack. Enjoy warm. The bannock will keep soft up to 24 hours, wrapped in kitchen towel and sealed in a polythene bag. It will freeze up to two months.

Also try adding to Maureen's recipe:

Spelt instead of wheat flour
Vary the quantities of flours to taste
1 tsp chopped fresh herbs
1 tbsp blitzed sunflower or pumpkin seeds
2 tsp poppy or sesame seeds
A handful of chopped walnuts

To make a meal with this bannock, stir into the mix grated cheese and chopped tomato, or wholegrain mustard and ham, smoked sausage or smoked salmon – what will you try in yours?

Scones and Pancakes

Victoria's beremeal blinis

Orcadian food writer Alan Bichan enjoyed Victoria Fairnie's beremeal blinis so much he persuaded her to share the recipe. She says 'I use equal amounts of beremeal and flour but, if you prefer a milder flavour, use four ounces of flour and two of beremeal. Serve with smoked salmon, crème fraiche and a little lemon zest.'

Makes 30 to 40 blinis

15g (½oz) butter
200ml (7fl oz) milk
85g (3oz) beremeal
85g (3oz) plain flour
½ tsp baking powder
Pinch of salt
½ tsp caster sugar
1 large egg, white and yolk separated
Butter for greasing

Put the butter and milk into a saucepan. Heat gently till the butter melts and the milk is lukewarm. Place the beremeal, flour, baking powder, salt, sugar and egg yolk in a food processor. Add the milk and process till smooth. Transfer to a bowl, cover with clingfilm and leave at room temperature for one hour. Whisk the egg white stiffly, stir one tablespoon into the mixture then gently fold in the remainder. Heat a heavy frying pan on medium and grease lightly with butter. Drop teaspoons of the batter onto the pan and cook for about one minute on each side. Keep warm in a low oven or allow to cool on a wire tray.

Salt-fish blinis with nori and crab

Sam Britten creates recipes using his experience as a chef working in top restaurants in this country and abroad. This dish will certainly impress all who come to dine at your table!

175ml (6 fl oz) milk
85g (3oz) beremeal
45g (1½oz) plain flour
1 x 7g (¼oz) pkt dried yeast
1 egg, white and yolk separated
100g (3½oz) of firm white fish, for example pollack
3 tsp salt
1 tbsp chopped parsley

To serve:
100g (3½oz) white crab meat
Vinegar – white wine, cider or elderflower
Salt and ground black pepper
3 sheets of nori seaweed, finely shredded

Rub the white fish with salt and marinate for 40 minutes. Poach in simmering water till tender. Drain and beat the fish till the flesh breaks into fibres. Sift the flours into a bowl. Heat the milk to blood-heat and mix with the flour and yeast. Cover and leave in a warm place for 1 to 2 hours to ferment. Fold in the egg-yolk, fish and parsley. Whisk the egg white to soft peaks and fold in. Fry over a medium heat in a little rapeseed or sunflower oil till golden on each side. Cool on a wire rack. Season the white crab meat with a little vinegar, salt and pepper. Top each pancake with crab meat, garnish with shredded nori and serve.

Norn beremeal drop scones

The Edinburgh restaurant Norn promotes the best of Scottish produce. Chef patron Scott Smith: 'Here is our recipe for beremeal drop scones. Really, really simple, but – awesome flavour from the beremeal. You can also add grated cheese or just serve them with bacon or ice cream.'

Makes 8 to 10 drop scones

115g (4oz) beremeal
15g (½oz) sugar
½ tsp bicarbonate of soda
1 egg
Pinch of salt
Milk to mix
Butter to cook

Sift the dry ingredients into a bowl and make a well in the centre. Add the egg and slowly beat in the milk to make a batter the consistency of thick cream. Heat a girdle or frying pan on medium to high heat and add a little butter. Drop a large dessertspoon of batter into the pan and cook for about 3 minutes, flip over and repeat on the other side. Remove from the pan and serve hot with bacon or ice cream, or Norn style: drizzle with birch syrup and serve.

Beremeal tattie scones

Floury tatties make lighter scones. Don't roll too thinly.

175g (6oz) mashed potato
10g (¼oz) melted butter
60g (2oz) beremeal
Pinch of salt

Heat a girdle or thick-bottomed frying pan on a low to medium heat. Mix the potato, beremeal and salt in a bowl, add melted butter and mix together. Turn the dough onto a board, dust with beremeal and knead

Scones on the girdle

till smooth. Cut in two and roll each half into a circle 5mm (¼in) thick and cut into quarters. Make sure the pan is not too hot by dusting with beremeal. If it burns reduce the heat. Bake the scones 3 to 4 minutes on each side till lightly browned. Cool in a clean tea towel on a wire tray. Enjoy freshly baked. Cut smaller triangles to serve with dips.

Try adding:

Cracked black pepper
Blitzed pumpkin seeds
Dried kelp or other seaweed

Joan's beremeal flatbreads

From my Orcadian friend, Joan: 'An easy recipe I made today for lunch – a beremeal flatbread. Serve with soup, with cheese and pickle or rolled with a filling. I find they work with anything spicy like coronation chicken, curries and tagines.'

Makes 4 or 5 flatbreads

200g (7oz) beremeal
Pinch of salt
100ml (3½ fl oz) water
2 tbsp oil (I used sunflower)

Sift the beremeal and salt into a bowl, stir in the water and oil. Turn onto a board dusted with beremeal and knead for 5 minutes. Roll out thinly into circles of whatever size you want. Heat a frying pan to medium heat, drizzle with a very little oil and cook the flatbreads for 3 minutes on each side. Keep warm in a clean tea towel on a wire rack.

Kirsty Aim's beremeal and cheese scones

Kirsty is the head cook at the Judith Glue Real Food Café in Kirkwall. She leads a small team who make good honest homemade food with local Orkney produce and has gained quite a reputation for her famous apple pies and cheese scones. This is her scone recipe.

Makes 12 large scones

350g (12oz) self-raising flour
350g (12oz) beremeal
2 tsp baking powder
1 tsp salt
1 tsp mustard powder
3 tbsp vegetable oil
300ml (10 fl oz) milk
2 eggs, beaten
225g (8oz) Orkney cheddar, grated

Heat the oven to 180ºC/fan 160ºC/350ºF/gas 4. Sift the dry ingredients into a bowl, make a well in the centre, add the eggs, milk, oil and cheese. Mix to a soft dough, adding more milk if needed. Do not over-mix. Turn onto a floured board, dust with flour and pat out to 2.5cm (1in) thickness. Cut with a 6cm (2½in) cutter and place on a floured baking tray. Dust the top with a little beremeal and grated cheese. Bake for 10 to 15 minutes till golden. Cool on a wire rack and enjoy warm and freshly baked.

Seeded walnut beremeal 'bannock-scone'

A quick 'throw in the oven' bake to feed unexpected guests.

Makes 8 wedges (farls)

100g (3½oz) beremeal
115g (4oz) self-raising flour
10g (2 tsp) baking powder
45g (1½oz) butter or margarine
1.2g (¼ tsp) sea salt
1.2g (¼ tsp) sugar
30g (1oz) chopped walnuts
30g (1oz) sunflower and 30g (1oz) pumpkin seeds, blitzed
2 large eggs, beaten
Milk to mix

Heat the oven to 200ºC/fan180ºC/400ºF/gas 6. Oil a baking tray. Sift the dry ingredients into a bowl. Rub in the butter. Add the salt, sugar, nuts and seeds. Mix to a stiff dough with the beaten egg and milk. Turn onto the tray, dust with beremeal and flatten into a round 3cm (1¼in) thick. Mark into 8 triangles. Bake for 12 to 15 minutes till firm. Cool on the tray. Serve the triangles (called farls) warm with cheese.

*A plate of beremeal scones. Made from a round
bannock cut into quarters these are often called farls*

Margaret's bacon and onion muffins

Another recipe by Margaret Phillips: 'I fancied making
a savoury muffin and used what I had to hand, but I
had no-one to taste. Caron and Ian at the local Palace
Stores obliged and gave them full marks!'

Makes 12 large muffins

100g (3½oz) streaky bacon (smoked or unsmoked),
    chopped
60g (2oz) onion, finely chopped
115g (4oz) beremeal
115g (4oz) spelt flour
2 tsp baking powder

½ tsp bicarbonate of soda
1 level tsp salt
Ground black pepper
2 large eggs, beaten
240ml (8fl oz) milk
120ml (4fl oz) natural yoghurt
60ml (4 tbsp) olive oil
1 dessertspoon runny honey
100g (3½oz) Orkney cheddar, grated

Heat the oven to 200°C/fan180°C/400°F/gas 6. Arrange 12 muffin cases on a tray. Sweat the bacon and onion over low heat to soften. Drain on kitchen towel. Sift the dry ingredients into a bowl. Mix the eggs, milk, yoghurt, oil and honey and add to the dry ingredients, along with the bacon, onion and grated cheese. Mix together and divide between the muffin cases. Bake for 15 to 20 minutes till risen and golden. Eat warm and freshly baked. You can keep them in the fridge for up to four days and re-heat before serving.

Bakehouse bere and raisin scones

A recipe by baker Jan Haetzer, who says 'Polenta makes a golden crusted scone with a moist fruity interior!'

Makes 8 scones 7.5cm (3in) in diameter

225g (8oz) plain flour
60g (2oz) beremeal
Pinch of salt
2½ tsp baking powder

50g (1¾oz) caster sugar
75g (2½oz) butter
60g (2oz) raisins
1 egg
150ml (5fl oz) milk
Polenta to roll out

Heat the oven to 220ºC/fan 200ºC/425ºF/gas 7. Oil a baking tray. Sift the dry ingredients into a bowl. Rub in the butter, then stir in the raisins. Beat the egg and milk together and mix into the dry ingredients to make a dough. Dust a board with polenta, turn out the dough and dust the top with polenta. Lightly roll to 2cm (¾in) thickness. Use a fluted scone-cutter (if possible) to cut out the scones. Knead the scraps and repeat. Dust again with polenta and place on the tray. Bake for 10 to 12 minutes till golden. Cool on a wire tray. Serve warm with plenty of butter. These scones keep fresh for 24 hours in a sealed container and freeze well.

Breads

David Hoyle, chef patron of Findhorn's Bakehouse Café, is a champion of Orkney beremeal. He introduced me to Jan Haetzer, the baker who supplied the previous recipe for Bakehouse bere and raisin scones. Jan is from an area of Germany near Leipzig. He was inspired to bake by spending time in a bakery opposite the family's farm there, and realised his ambition eight years ago when he moved to Scotland. These are two of his bread recipes.

If possible bake both breads with the oven fan turned off.

Bakehouse beremeal sourdough bread

Makes 2 x 900g (2lb) loaves

Sourdough starter

A sourdough starter takes 10 days to make. Put it in a large glass jar or bowl with a lid. Scald this with boiling water to sterilise.

Day 1. Mix 100g (3½oz) rye flour with 100ml (3½fl oz) water. Cover. Leave for three days at room temperature.

Day 3. Mix in 50g (1¾oz) rye flour and 50ml (1¾fl oz) water. Cover and leave two days.

Repeat this step twice more.

Day 10. The starter should have a pleasant, slightly sour smell. Discard if it has gone mouldy or smells vinegary.

If you keep the starter alive, it will last years. Keep the starter in a container at room temperature and feed with flour and water as you use it or every two to three days. Alternatively, keep the starter in the fridge for up to four weeks and feed each week with flour and water. The starter will keep frozen or dried and crumbed and stored in a sealed jar. Both are re-activated by adding flour and water and returning to room temperature. Readymade cultures are available to buy.

Making the bread

Day 1:

60g (2oz) sourdough starter
320g (11oz) beremeal
400ml (14fl oz) cold water

Mix together using a dough-hook or K-beater for 5 minutes on slow speed or mix using your hand or a wooden spoon to a smooth dough. Cover and leave at room temperature overnight.

Day 2, add:

320g (11oz) beremeal
320g (11oz) rye flour
400ml (14fl oz) warm water
10g (2 tsp) salt

Add the above ingredients to the initial dough and mix with the dough hook or K-beater on slow speed for 20 minutes. If hand-mixing, knead the dry ingredients into the initial dough on a clean work surface or in a large bowl for about 20 to 30 minutes, or till the dough is smooth and stretchy. Oil the loaf tins well. Divide the dough in half and place in each tin. Smooth with a wet hand. Cover with a damp tea-towel and prove for 1½ hours in a warm place (about 35ºC) till risen, firm to touch and approximately doubled in size. This can take up to 2½ hours in colder weather. Heat the oven to 230ºC/fan 210ºC/450ºF/gas 8. Bake for 20 minutes, cover with greaseproof paper and bake a further 30 minutes. The bread is ready when it comes out of the tin cleanly, the sides are brown and it sounds hollow when knocked on the base with the knuckles. Cool on a wire rack. Store in a sealed container or in the fridge for up to 1 week. The loaves will freeze for up to 2 months.

Beremeal and spelt baguette

1 kg (2¼lb) wholemeal spelt flour
500g (1lb 2oz) beremeal
30g (1oz) salt
1 litre (1¾ pt) cold water
50g (1¾oz) fresh yeast or 1.5 x 7g packets of easy-bake yeast
1 tbsp malt extract

The day before you want to bake, sift the dry ingredients into a mixing bowl. Use a dough hook or K-beater to mix in the water and malt on slow speed for 7 minutes, then 3 minutes on fast speed. If mixing by hand, combine the ingredients using a wooden spoon and then knead by hand till smooth. Turn the dough onto a board dusted with a mixture of spelt flour and beremeal. Divide it into 10 equal pieces with a bakery cutter or long-bladed sharp knife. Roll each into a baguette shape about 25cm (10in) long and put on baking trays and keep in the fridge overnight. The next day, take them out of the fridge and leave to reach room temperature. Heat the oven to 230ºC/fan 210ºC/450ºF/gas 8. Bake for 13 to 17 minutes till golden and baked through. Test by knocking the base of a baguette with your knuckles; if it sounds hollow, the bread is ready. Cool on a wire tray. Serve warm and freshly baked.

Beremeal treacle soda bread

Two attempts at this recipe using beremeal were rejected by my Irish neighbour Aubrey. Third time lucky!

Makes a round 20cm (8in) across

115g (4oz) self-raising flour
85g (3oz) beremeal
¼ tsp salt
1 tsp baking powder
½ tsp bicarbonate of soda
Handful of raisins (optional)
1 tsp ground ginger
30g (1oz) each of black treacle and honey
45g (1½oz) melted butter
150ml (5fl oz) buttermilk or milk soured with 1 tbsp of fresh lemon juice

Heat the oven to 190°C/fan 170°C/375°F/gas 5. Oil a baking tray. Sift the dry ingredients into a mixing bowl. Melt the treacle, honey and butter together, but do not boil. Pour into the dry ingredients. Mix with buttermilk or soured milk to make a soft dropping consistency. Scrape onto the middle of the baking tray and dust with beremeal. Shape into a round and cut a deep cross in the top with a long-bladed knife. Bake for 25 minutes till risen and firm. Cool on a wire tray. This is best eaten freshly baked.

Biscuits

Beremeal and tattie crackerbread

When meal was scarce, Orcadian housewives added tatties to their bannocks. I turned their recipe idea into biscuits!

Makes 12 thin fingers 2.5 x 12cm (1 x 5in)

50g (1¾oz) beremeal
5ml (1 tsp) oil
85g (3oz) smooth mashed potato

Turn on the oven at 180ºC/fan 160ºC/350ºF/gas 4. Oil a baking tray. Put the ingredients in a bowl and mix together. Divide in two. Roll each half between two sheets of clingfilm to a thin strip 12cm (5in) wide. Remove the top film, lift the dough on the lower sheet over the baking tray, invert onto the tray and remove the film. Cut across the strip in 2.5cm (1in) intervals to make fingers. Bake for 12 to 15 minutes till crisp. Watch carefully as they burn easily. Cool on a wire tray and store in an airtight container.

Serve with dips and cheeses.

Norn beremeal crackers

At one time the people of Orkney spoke Norn, a variant of Old Norse. Norn is also the name of an Edinburgh restaurant where chef patron Scott Smith bakes with beremeal.

Makes 50 to 60 crackers

125g (4½oz) strong bread flour
145g (generous 5oz) beremeal
40g (1¼oz) rye flour
2.5g (½ tsp) salt
30ml (2 tbsp) water
7.5g (¼oz) fresh yeast or 2.5g (½ tsp) easy-bake yeast
140ml (4½fl oz) whole milk
Rapeseed oil

Sift the dry ingredients into a bowl. Whisk the water, yeast and milk together. Mix both to make a dough, cover and rest in a warm place for 1 hour. Knead the dough for 10 minutes to develop the gluten. Cover and rest for 45 minutes. Heat the oven to 220ºC/fan 200ºC/400ºF/gas 6. Oil and line a baking tray with non-stick baking parchment and brush this with oil to prevent the crackers sticking. Roll out the dough very thinly to 2mm. Cut into any size or shape you wish and place on the baking tray. Drizzle with a little rapeseed oil and bake for about 10–15 minutes till dark golden brown. Remove from the oven and cool on a wire rack. These will store for up to 10 days in a sealed container.

Seeded bere and oat crispbread

This is based on a Norwegian crispbread recipe.

Makes 30 crackers 3.5 x 6cm (1½ x 2½in)

50g (1¾oz) beremeal
50g (1¾oz) flaked oats

30g (1oz) sunflower seeds, blitzed
30g (1oz) pumpkin seeds, blitzed
2g (¼ tsp) fine sea salt
100ml (3½fl oz) tepid water

Heat the oven to 180ºC/fan 160ºC/350ºF/gas 4. Oil a non-stick baking tray. Put the beremeal, oats, seeds and salt into a bowl and mix with water to a smooth dough, adding more water if needed. Knead in the bowl till smooth. Roll as thinly as the seeds will allow between two sheets of clingfilm. Peel off the top sheet then lift the bottom sheet and invert over the baking tray to deposit the dough flat on it. This is not so easy to do, but if the dough breaks, press together and roll flat once more. Mark into rectangles with a long-bladed knife. Bake for 15 minutes. Open the oven door to let steam out, then bake a further 10 to 12 minutes or till crisp. Cool a little on the tray. Loosen the crisp bread with the blade of a palette knife and lift onto a wire rack. Break at the marked rectangles when cold. Store in an airtight tin.

Try adding sesame, linseed or poppy seeds to the mix.

Margaret Phillips' beremeal water biscuits

Another recipe from Margaret Phillips. 'This recipe was transcribed from my Aunt Janet Sutherland's recipe book. Born in Shetland in 1910, she was a housekeeper/cook all her working life and during the war worked for influential business people in Glasgow and other places.'

Makes around 18 biscuits if using a 7.5cm (3in) cutter

75g (3oz) strong white flour
25g (1oz) beremeal
10g (½oz) lard
1 level tsp salt
75ml (2½fl oz) water

Heat the oven to 180ºC/fan 160ºC/350ºF/gas 4. Sift the dry ingredients and rub in the lard. Mix to a dough with water. Use plenty of flour on the rolling pin and board and roll into a rectangle. Pat the entire surface with the back of a wooden spoon. Fold the dough into three. Seal the open edges by pressing firmly with the rolling pin.

Repeat the folding process twice more. Roll out very thinly. Cut into rounds and prick all over with a fork, making neat patterns. Place on lightly floured oven trays. Bake for 8 minutes, change the upper and lower trays around and bake a further 8 minutes. Keep a careful watch, as they will easily burn. Cool on a wire rack. Store in an airtight container.

Beremeal oatcakes

Nutty and delicious, beremeal oatcakes complement Orkney cheeses perfectly.

Makes 24 x 6cm (2½in) oatcakes

115g (4oz) medium oatmeal
85g (3oz) beremeal
½ tsp sea salt

Generous pinch of bicarbonate of soda
30ml (2 tbsp) sunflower oil
Tepid water to mix

Heat the oven to 180°C/fan 160°C/350°F/gas 4. Put the meals, salt and bicarbonate of soda in a bowl and add the oil. Mix to a clean pliable dough with tepid water. Turn onto a beremeal-dusted work surface and knead together. Roll out thinly to 3mm (1/8in), dusting with beremeal to prevent sticking. Cut into 6cm (2½in) rounds. Bake for 12 to 15 minutes till crisp. Watch carefully as they burn easily. Cool on a wire tray and store in an airtight tin.

Alternatively, halve the dough and roll each half thinly into rounds. Cut across the centre to make four or six triangles. Bake as above. The round is called a bannock, the triangles farls.

For additional flavour add:

1 tsp roughly ground black pepper
1 dessertspoon crushed pumpkin seeds
1 tsp poppy seeds

Beremeal and oat digestives

Not overly sweet, these are a wheat-free crumbly eat with tea, coffee or Orkney cheeses.

Makes 22 5cm (2in) biscuits

100g (3½oz) beremeal
100g (3½oz) fine or medium oatmeal

30g (1oz) light soft brown sugar
75g (2½oz) butter or margarine
Tepid water to mix

Heat the oven to 180ºC/fan 160ºC/350ºF/gas 4. Put the meals in a bowl. Cream the butter and sugar till light. Mix in the meal on slow speed or by hand using a wooden spoon, adding a little tepid water to make a clean pliable dough. Turn the dough onto a beremeal-dusted work surface and knead smooth. Roll out to 5mm (¼in) thickness. Cut with a 5cm (2in) round fluted cutter and put onto a baking tray. Prick with a fork once in the middle of each biscuit. Bake for 12 to 15 minutes till coloured. Watch carefully because the biscuits burn easily. Cool on a wire tray and store in an airtight tin.

As a variation, add:

1 tsp cracked black pepper
Blitzed sunflower seeds
Poppy seeds

Beremeal and spelt shortbread

These two ancient grains make a wholesome short-bread packed with flavour.

Makes 18 x 5cm (2in) biscuits

60g (2oz) butter
45g (1½oz) caster sugar
60g (2oz) spelt flour

60g (2oz) beremeal
10g (¼oz) cornflour

Heat the oven to 180ºC/fan 160ºC/350ºF/gas 4. Cream the butter and sugar till light and sift in the flours. Stir them into the creamed mixture, adding a little tepid water if needed to make a clean, pliable dough. Knead lightly then roll out on a floured board to 5mm (¼in) thickness. Cut into 5cm (2in) round biscuits; I like to use a cutter with crimped edges for sweet biscuits. Bake for 15 to 20 minutes till golden and crisp through. Dust with caster sugar when warm, then cool on a wire rack. Store in an airtight tin.

Honey crunchies

From an old family recipe mother called 'waffle biscuits'.

Makes 18 biscuits

75g (2½oz) beremeal
1 tsp baking powder
60g (2oz) rolled oats
45g (1½oz) butter
30g (1oz) soft brown sugar
30g (1oz) honey
¼ tsp bicarbonate of soda

Heat the oven to 180ºC/fan 160ºC/350ºF/gas 4. Sift the beremeal and baking powder into a bowl, add the oats. Melt the butter, sugar and honey in a pan, then beat in bicarbonate of soda till the mixture is frothing.

Mix in the dry ingredients. Put a teaspoon at a time onto non-stick baking trays, leaving room for the biscuits to spread. Make a waffle pattern on each biscuit with back of a fork. Bake for 8 to 9 minutes. Watch carefully as they burn easily. Leave to set on the tray, use a palette knife to loosen, and cool completely on a wire tray. Store in a sealed container.

Ginger almond Florentines

A local Scottish Women's Institute recently entered a competition to bake a 'Victorian afternoon tea'. Each member contributed. My friend Valerie made Florentines so I volunteered to help. Experience taught us that we should have the ingredients weighed out before mixing and bake in batches to avoid burnt offerings!

Makes 30 biscuits

Melt:
120ml (4fl oz) double cream
60g (2oz) butter
30g (1oz) honey
85g (3oz) caster sugar

Mix:
45g (1½oz) beremeal
10g (¼oz) plain flour
1 tsp ground ginger
30g (1oz) mixed peel
150g (5oz) flaked almonds

To coat:
200g (7oz) dark chocolate (if liked)

Heat the oven to 180ºC/fan 160ºC/350ºF/gas 4. Oil 2 baking trays. Put the cream, butter, honey and sugar in a pan and bring to the boil, stirring continuously. Remove from the heat and stir in the rest of the ingredients. Rest the mixture for 2 or 3 minutes. Drop a teaspoon at a time onto the trays, leaving about 7.5cm (3in) space between. Flatten each using the back of a teaspoon dipped in hot water. Bake for 8 to 10 minutes, till golden brown and bubbling. Remove from the oven and cool on the tray for 2 or 3 minutes to set. Use the flat blade of a palette knife to lift onto a wire tray. Repeat with the rest of the mixture.

If coating, melt the chocolate in a bowl over a pan of simmering water. Remove from the heat and stir till the chocolate begins to thicken. Use a small palette knife to spread the flat side of the Florentines with dark chocolate. Lay on a tray, chocolate side up, to set. Store in an airtight container.

Fruit Loaves

Beremeal Earl Grey tea bread

A close friend enjoys drinking numerous cups of Earl Grey tea over cake and chat. On one such occasion he was called away unexpectedly. What to do with the tea? I made a cake.

Makes 2 x 450g (1lb) loaves

Soak overnight:
400g (14oz) dried fruit
300ml (10fl oz) hot Earl Grey tea

Add:
115g (4oz) beremeal
140g (5oz) self-raising flour
2 tsp baking powder
1 tsp mixed spice
115g (4oz) soft brown sugar
85g (3oz) melted butter or sunflower oil
1 large egg, beaten
Grated rind of half a lemon

To glaze:
1 tbsp honey
1 tbsp lemon juice

Heat the oven to 180°C/fan 160°C/350°F/gas 4. Oil and line the loaf tins. Sift the beremeal, flour, baking powder and spice into the soaked fruit. Add the sugar, butter or oil, egg and lemon. Stir together. Pour into the loaf tins. Bake for 35 to 40 minutes till risen and firm and the point of a skewer inserted in the middle comes out cleanly. Glaze with honey and lemon juice and cool in the tin. Keep at least one day before cutting. Store, foil-wrapped, in an airtight container. This will freeze for up to two months.

As a variation my friend Joan used a 'tea' made by infusing mulled wine 'tea'-bags – delicious.

Enjoying a cup of tea and some beremeal baking at the North Ronaldsay Harvest Home

Fruity beremeal banana loaf cake

Another recipe by Margaret Phillips. 'Make sure to use really ripe bananas. Under-ripe fruit does not give such good results.'

Makes 2 x 450g (1lb) loaves

125g (4½oz) butter
250g (9oz) caster or soft brown sugar
4 ripe bananas, peeled and mashed
2 eggs, beaten
250g (9oz) beremeal
3 tsp baking powder
100g (3½oz) sultanas
100g (3½oz) chopped crystallised ginger

Heat the oven to 160ºC/fan 140ºC/325ºF/gas 3. Oil and line the loaf tins. Cream the butter and sugar well. Beat in the mashed banana and eggs. Sift in the beremeal and baking powder and fold in. Stir in the sultanas and ginger. Divide between the loaf tins. Bake for 50 minutes till risen and when the point of a skewer inserted in the middle comes out cleanly. Cool a little in the tin, and then on a wire tray. Store, foil-wrapped, in an airtight container. This will freeze for up to two months.

Seeded orange and date chai tea loaf

Chai tea adds a note of warm Indian spice.

Makes 2 x 450g (1lb) loaves

To soak:
115g (4oz) sultanas
115g (4oz) chopped dates
600ml (1pt) vanilla chai tea

Add:
115g (4oz) soft brown sugar

115g (4oz) honey
175g (4oz) orange marmalade
225g (8oz) beremeal
225g (8oz) spelt flour
6 tsp baking powder
30g (1oz) pumpkin seeds, blitzed or roughly chopped
30g (1oz) sunflower seeds, blitzed or roughly chopped

Heat the oven to 160ºC/fan 140ºC/325ºF/gas 3. Oil and line the loaf tins. Soak the sultanas and dates in hot tea for at least 1 hour or overnight. Add the sugar, honey and marmalade and stir together. Sift the beremeal, flour and baking powder into a bowl and add the pumpkin and sunflower seeds. Stir in the fruit mixture to make a soft dropping consistency. Add more tea if needed. Pour into the loaf tins. Bake for 45 to 50 minutes till risen and when the point of a skewer inserted in the middle comes out cleanly. Cool a little in the tin, and then on a wire tray. Store, foil-wrapped, in an airtight container. This will freeze for up to two months.

Beremeal, banana, chocolate and walnut

Caron and Ian Smith-Crisp own and run Birsay's Palace Stores where Caron's home-bakes are in demand. This is her recipe. 'My mum taught me to bake, and since moving to Orkney I have discovered that beremeal can be used with success in virtually any recipe.'

Makes 2 x 450g (1lb) loaves

175g (6oz) golden caster sugar
115g (4oz) soft butter
100g (3½oz) beremeal
60g (2oz) self-raising flour
2 large or 3 small bananas, sliced
125g (4½oz) dark chocolate chunks, half of them
    chopped into smaller pieces
85g (3oz) walnuts, chopped
2 eggs, beaten
4 tbsp milk
1 tsp bicarbonate of soda

Heat the oven to 180ºC/fan 160ºC/350ºF/gas 4. Oil and line the loaf tins. Put all the ingredients in a food mixer or processor. Mix on slow speed for 30 seconds then on medium till well blended and the bananas are squashed. Alternatively mix by hand using a wooden spoon. Don't worry if the mix looks curdled. Divide between the loaf tins. Bake for 25 to 30 minutes till risen and firm. Do not over-bake. Cool in the tin. Store, foil-wrapped, in a sealed container for up to one week. This will freeze for up to two months.

Cakes

Highland 'Park'-in

My late father preferred blended whiskies until, aged 90, he savoured a glass of Highland Park malt. 'Why did no-one tell me what a fine whisky this is?' he exclaimed. Later, en route to Orkney, I offered to buy a bottle at the distillery 'A bottle!' he exclaimed, 'Make it a case!' Dad was also partial to a moist gingerbread, so I combined the two in this recipe.

Makes a 23cm (9in) square tin

115g (4oz) beremeal
115g (4oz) self-raising flour
2 tsp baking powder
1 level tsp bicarbonate of soda
1 tsp ground ginger
½ tsp ground mixed spice
115g (4oz) medium oatmeal
85g (3oz) butter
85g (3oz) soft brown sugar
60g (2oz) black treacle
85g (3oz) golden syrup
1 large egg, beaten
150ml (¼ pt) milk to mix
2 tbsp hot water
2 tbsp Highland Park whisky

Heat the oven to 160ºC/fan 140ºC/325ºF/gas 3. Oil and line the tin. Sift the beremeal, flour, baking powder, bicarbonate of soda and spices into a bowl and add the oatmeal. Melt the butter, sugar, treacle and syrup and pour into the dry ingredients. Mix to a soft dropping consistency with the egg and milk, then beat in the hot water. Pour into the tin. Bake for 30 minutes till risen and when the point of a skewer inserted in the middle comes out cleanly. Cool a little and drizzle with whisky. Leave to cool in the tin. Mature for 24 hours before cutting. Serve cut into generous squares. Store, foil-wrapped, in an airtight tin.

Also try:
As a dessert with stewed rhubarb and thick cream
With a light cream cheese frosting

Ginger 'bere' meal cake

This was test-baked by my Orcadian friend Joan: 'I made the cake today and it was delicious.'

Makes a tin 23 x 15cm (9 x 6in) x 6cm (2½in) deep

175g (6oz) self-raising flour
175g (6oz) beremeal
2 level tsp baking powder
2 tsp ground ginger
240ml (8fl oz) ginger or similar beer
115g (4oz) melted butter
115g (4oz) runny honey or golden syrup
115g (4oz) soft brown sugar
3 large eggs

For the topping:
30g (1oz) dark brown sugar to finish

Heat the oven to 150ºC/fan 130ºC/300ºF/gas 2. Oil and line a baking tin. Beat all the ingredients together and pour into the tin. Sprinkle sugar on top. Bake for 55 minutes till firm and set. Cool in the tin. Store, foil-wrapped, in an airtight container for up to 1 week. This will freeze for up to two months.

Carol Wilson's beremeal fruit cake

Author, broadcaster and food writer Carol Wilson has a particular interest in food history, so was pleased to create this special recipe. We have been friends for many years. Carol makes amazing cakes! She says 'This is a lovely moist cake, which tastes even better if kept in an airtight tin for a day or so before eating. I like to ring the changes using different-flavoured teas – orange, spice, ginger or apple are good choices.'

Makes a 20cm (8in) round cake

450g (1lb) mixed dried fruits: currants, raisins, sultanas
125ml (4fl oz) hot black tea
200g (7oz) unsalted butter
200g (7oz) light muscovado sugar
300g (10oz) self-raising flour
100g (4oz) beremeal
2 tsp baking powder
1 tsp mixed spice
1 tbsp orange or ginger marmalade
4 eggs, beaten

Soak the dried fruits in the tea overnight. Heat the oven to 180ºC/fan 160ºC/350ºF/gas 4. Oil and line a cake tin. Beat together the butter and sugar in a mixing bowl until light. Beat in the eggs, a little at a time, adding a spoonful of flour if the mixture curdles. Sift in the dry ingredients and mix well. Stir in the soaked fruit and marmalade. Pour into the tin and level the surface. Bake for 20 minutes. Reduce the oven temperature to 150ºC/fan 130ºC/300ºF/gas 2 and continue baking for 1 to 1½ hours until cooked through. Cool completely in the tin. Store the cake in an airtight container.

Caron's cider apple cake

Another recipe from Caron Smith-Crisp. Before she and her husband Ian took over Birsay's Palace Stores in January 2015 they lived in the Yorkshire Dales. Caron says: 'In Yorkshire there is a tradition of eating fruit cake with Wensleydale cheese. This is a moist sweet cake but cheese gives a wonderful sour savoury note – especially delicious!'

Makes 2 x 450g (1lb) loaves

150ml (5fl oz) cider
60g (2oz) dried mixed fruit
115g (4oz) butter or margarine
175g (6oz) unrefined caster sugar
2 eggs, beaten
140g (5oz) beremeal
60g (2oz) self-raising flour
50g (1¾oz) ground almonds

½ tsp bicarbonate of soda
½ tsp cinnamon
2 eating apples, washed and grated
85g (3oz) crumbly cheese (optional)

To glaze:
Apricot jam or honey

Heat the oven to 170ºC/fan 150ºC/325ºF/gas 3. Oil and line the tins. Put the fruit and cider into a bowl and microwave for 3 or 4 minutes or heat in a pan. Cream the butter and sugar till fluffy. Beat in the egg. Sift in the dry ingredients and stir together. Fold in the almonds, apple, fruit mix and cheese (if using). Divide the mixture between the tins. Bake for 30 to 35 minutes till risen, firm and golden. Cool in the tin. Glaze with warm jam or honey.

Also try:
Before baking, lay apple slices down the middle and sprinkle with crumbled cheese or flaked almonds, then bake as above.

Beremeal 'Black Forest' brownies

The name of this recipe is a contradiction in terms – there are no forests on Orkney! Another great bake by Margaret Phillips.

Makes a cake 20 x 30cm (8 x 12in)

180g (generous 6oz) unsalted butter
200g (7oz) dark chocolate, chopped

4 eggs
100g (3½oz) caster sugar
85g (3oz) soft brown sugar
150g (5½oz) beremeal
1 tsp baking powder
100g (3½oz) dried dark sour cherries
60g (2oz) flaked almonds
Icing sugar

Heat the oven to 160ºC/fan 140ºC/325ºF/gas 3. Oil and line a cake tin. Put the butter and chocolate into a bowl over a pan of simmering water and stir till melted. Remove from the heat to cool. Whisk the eggs and sugars till thick and creamy. Sift the beremeal and baking powder into the bowl and pour in the cooled chocolate mix. Fold together along with the cherries and almonds. Pour into the prepared tin and bake for 30 minutes till risen and softly set and when the point of a cocktail stick inserted in the middle is lightly sticky. Cool in the tin, dust with icing sugar, cut into squares and serve. Store up to 1 week in a sealed container. This can be frozen for up to two months.

Chocolate and beetroot beremeal brownies

Beetroot is tricky to handle, a few drops can make a large stain. This recipe is designed with that in mind!

Makes 1 20 x 30cm (8 x 12in) tin

250g (9oz) peeled and cooked beetroot (without vinegar)

100ml (3½fl oz) vegetable oil
60g (2oz) 70% cocoa dark chocolate
3 large eggs
115g (4oz) caster sugar
85g (3oz) honey (or agave syrup)
1 tsp vanilla essence
75g (2½oz) self-raising flour
75g (2½oz) beremeal
75g (2½oz) cocoa powder
1 tsp baking powder
Pinch of salt
60g (2oz) chopped walnuts

For the icing:
100g (3½oz) icing sugar
Beetroot juice
Fresh lemon juice

Heat the oven to 180°C/fan 160°C/350°F/gas 4. Oil and
line a cake tin. Blend the cooked beetroot till smooth.
Extract a teaspoon of juice and save to colour the icing,
pour the puree into a mixing bowl and beat in the oil.
Melt the chocolate in a bowl over a pan of simmering
water, stir till smooth and then remove from the heat
to cool. Whisk the eggs, sugar and honey till thick and
creamy. Stir in the vanilla and melted chocolate, then
fold into the beetroot mix. Sift the flour, beremeal,
cocoa and baking powder into the bowl, add walnuts
and fold in. Pour into the tin. Bake for 20 to 25 minutes
till firm but soft, and the tip of a wooden cocktail stick
remains sticky when removed. If the cake is still soft
in the middle, lay a sheet of baking paper over the top

of the tin to prevent the cake darkening too much on top till cooked through. Cool in the tin. Mix the icing ingredients together to drizzle over the cake. Leave to set. Serve cut into squares.

Chewy date and apple bere bars

This is a sustaining snack and can even provide 'breakfast on the go'.

Makes 24 bars in a tin 20 x 30cm (8 x 12in)

115g (4oz) butter or margarine
60g (2oz) soft brown sugar
60g (2oz) syrup
115g (4oz) chopped dates
115g (4oz) rolled oats
115g (4oz) beremeal
2 eating apples, peeled and grated
60g (2oz) mixed seeds: pumpkin, sunflower, linseed, sesame
115g (4oz) dark chocolate (optional)

Heat the oven to 180°C/fan 160°C/350°F/gas 4. Oil the baking tin. Heat the margarine, sugar and syrup to melt together, mix in the dates and stir to soften. Stir in the oats, beremeal, apple and seeds. Spoon into the tin and spread evenly. Bake for 20 minutes till firm. Cool in the tin. Cut into fingers while warm.

To coat them, melt chocolate in a pan over simmering water. Stir to cool the chocolate and spread over the cooled bars. Keep in a sealed container.

Fairtrade banana beremeal muffins

It is fitting that the Orkney Islands are designated a Fairtrade area, because generations of Orcadians sailed the seas to trade with the very lands supported by the scheme. The Orkney Fairtrade Group achieved Fairtrade Zone status in February 2014 and have worked hard to sustain this ever since. Their slogan is 'Buy Fair and Buy Local'.

> Fairtrade, Fairtrade, twa wurds displayed
> Fur goods that's either grown or made,
> Twa wurds that tell us at a stroke
> That wur supportan fermeen folk,
> That the producers' price is fair
> An' middlemen don't tak their share.
> The cash we pay fur goods or food
> Returns to source an's daein' guid,
> So choose Fairtrade tae source yur meal
> Then fermers get a better deal.
> Buy local, fresh wi' low food miles,
> Buy local and support the isles!
> Buy local, and each pound you spend
> Stays here gaan roond fur months on end!

From 'Fair Trade from an Orkney perspective',
by Councillor Harvey Johnston.

This recipe for Fairtrade muffins is from Gill Smee, chair of Orkney Fairtrade Group, who devised it to use up squashed bananas left over from the fun Fairtrade banana tower building competition, a popular feature on their stand at food events.

Makes 9 muffins

75g (3oz) beremeal
75g (3oz) self-raising flour
2 tsp baking powder
60g (2oz) Fairtrade Demerara sugar
50g (1¾oz) Fairtrade chocolate, chopped into small
    pieces
60g (2oz) melted butter
1 ripe Fairtrade banana, mashed
1 large egg, beaten
100ml (3½fl oz) milk

Heat the oven to 190°C/fan 170°C/375°F/gas 5. Put 9 muffin cases in a muffin tin. Sift the dry ingredients into a bowl. Add the sugar and chocolate and mix together. Stir in the mashed banana, egg and milk and spoon into the muffin cases. Bake for 12 to 15 minutes till firm and when the point of a knife or skewer inserted in the middle comes out cleanly. Cool on a wire tray. Enjoy freshly baked. These muffins keep in a sealed container for up to three days and freeze well.

Fairtrade beremeal very 'ginger' bread

Makes 2 x 450g (1lb) loaves

150g (5½oz) self-raising flour
150g (5½oz) beremeal
1 heaped tsp Fairtrade ground ginger
1 level tsp bicarbonate of soda
85g (3oz) butter
115g (4oz) Fairtrade soft brown sugar

60g (2oz) Fairtrade raisins
60g (2oz) black treacle
60g (2oz) golden syrup
1 large egg, beaten
150ml (5fl oz) milk
2 tbsp hot water

Heat the oven to 160ºC/fan 140ºC/325ºF/gas 3. Oil and line the loaf tins. Sift the flour, beremeal, ginger and bicarbonate of soda into a bowl. Rub in the butter and stir in the sugar and raisins. Melt the treacle and syrup and add to the mix along with the egg. Beat in the milk till smooth, followed by the hot water to make a soft, almost pouring consistency. Pour into the tins. Bake for 25 to 30 minutes till risen, firm and when the point of a skewer or knife inserted in the middle comes out cleanly. Cool in the tins. Keep for one day to mature before cutting. Store, foil-wrapped, in an airtight container. This freezes well.

Savoury Dishes

Yorkney puddings

Andy Mitchell came to Orkney in 1990, working first on North Ronaldsay at the bird observatory, then as RSPB warden on Rousay and Egilsay. Now semi-retired, he still shares his knowledge with birdwatchers at home in South Ronaldsay and on trips to Cuba. He writes: 'After hearing your request for recipes using beremeal, I had an idea! Yorkshire puddings! Last night we tried it made with half beremeal and half flour. Well, they were delicious! Not risen quite as much, but texture and taste just excellent.'

Makes 6 puddings in a standard 6cm (2½in) bun tray

50g (1¾oz) beremeal
50g (1¾oz) plain flour
60ml (2fl oz) milk
3 tbsp water
1 egg
Salt and ground black pepper
Oil – choose a fat or oil stable at high temperature

Sift the dry ingredients into a bowl and add the egg, milk, water, a pinch of salt and black pepper. Blend with an electric or hand whisk and let it stand for ten minutes. Heat the oven to 200ºC/fan 180ºC/400ºF/ gas 6. Put a dribble of oil into six cups in a suitable tray. Heat the oil in the oven till very hot. Whisk the mixture again to add more air, adding a little water if too thick – the batter should be the consistency of pouring single cream. Pour into a jug and then into each cup just shy of the top. Bake for about 30 minutes till risen and brown. Serve at once.

Spinach and hummus roulade

A recipe by food writer Rosemary Moon, who says: 'This sounds complicated but it isn't at all. You can make it with watercress instead of spinach, and add finely chopped or shredded salad vegetables to the hummus filling if you wish – spring onions and shredded peppers are particularly good.'

Makes a Swiss roll tin 23 x 30cm (9 x 12in)

100g (2½oz) small spinach leaves or watercress
2 tbsp beremeal
Salt and ground black pepper
4 large eggs
175–200g (6–7oz) hummus
Salad to serve

Heat the oven to 220ºC/fan 200ºC/425ºF/gas 7. Oil and line the tin with baking parchment. Finely chop the spinach in a food-processor, add beremeal with salt and pepper and mix together. Whisk the eggs for 5 minutes until thick and pale in colour. The whisk leaves a thick trail when the mixture is ready. Carefully fold in the spinach and beremeal. Turn the mixture into the prepared tin and spread evenly. Bake about 10 minutes. The baked roulade will be flatter; this is normal. Carefully turn the roulade out onto a cooling rack lined with baking parchment. Gently ease it off the parchment in which the mixture was baked, then roll up the roulade with the new parchment within the roll because it is to be unrolled again when cool to fill with the hummus. Roll from the long side for starter-sized pieces and from the short side for more substantial main course slices. Once cool, unroll, spread with hummus, then roll it up again without the baking parchment and slice. Serve with a leaf salad in the summer and a crunchy salad like a Waldorf in the winter.

Rosanna's beremeal pasta

I wondered if it was possible to make beremeal pasta, so I asked my Italian friend Rosanna to help. She makes fresh pasta to the recipe her mother and aunts used when she was a child. Her only modern concession is to use a food-processor in place of the traditional hand-mixing method. Rosanna comes from Borgotarra in the Italian province of Parma, famous for its porcini mushrooms and parmigiano cheese.

Makes enough for 4

85g (3oz) beremeal
175g (6oz) plain flour
1 level teaspoon salt
2 large fresh eggs
½ tbsp olive oil
½ tbsp water
Butter or olive oil

Sift the dry ingredients into the bowl of a food-processor or mixer. Add the eggs, oil and water. Mix in bursts of 30 seconds, switching on and off till the dough comes together cleanly. Turn onto a floured board and knead till smooth. Cut into 4 pieces. Roll each into a thin strip. Dust the top with beremeal and fold over in a loose flat roll. Use a sharp knife to shred thinly from end to end, then use your fingers to lift gently and shake to loosen the shreds of pasta. The dough can also be processed through a pasta machine. Toss into boiling salted water and simmer for 15 to 20 minutes. Drain through a sieve and return to the hot pan, tossing with butter or olive oil to coat the pasta. Serve in warm bowls topped with freshly grated parmesan or stir the parmesan through to melt before serving. Beremeal pasta is so full of flavour it needs nothing else. Make lasagne by cutting pasta sheets in rectangles instead of noodles. This pasta will keep for up to three days in the fridge and freezes up to two months.

Christopher Trotter's bere 'n' carrot fritters

Chef Christopher Trotter is a cookery writer and broadcaster, often heard on Radio Scotland's *Kitchen Café* and *Kitchen Garden*. 'The recipe is based on one from my series of four vegetable books. This is from the carrot book. These are delicious if deep-fried and make great snacks with Trotter's hot pepper jelly, and a glass or two of beer!'

Serves 4 to 6

1 tsp ground cumin
1 tsp ground coriander
¼ tsp turmeric
150g (5½oz) beremeal
125ml (generous 4fl oz) Skull Splitter beer or a light
    smooth ale
1 egg, lightly beaten
200g (7oz) carrots, grated
1 bunch spring onions, finely chopped
1 tbsp chopped mint
1 tsp sea salt
500ml (16fl oz) oil for deep fat frying

In a bowl mix the spices with the beremeal and make a batter with the beer and egg. Squeeze out excess moisture from the grated carrot and mix into the batter with the spring onions and mint. Season with sea salt; a sea salt with added seaweed works well. Drop rounded dessertspoons of mixture into hot oil (180°C/350°F) turning to brown all over; this takes about 5 minutes. Drain on kitchen towel and serve hot.

JACKS' 'berely' battered fish

Pierowall Fish Ltd, on the north isle of Westray, is owned and run by Kevin and Ann Rendall. On Wednesday and Saturday nights their 'JACKS' Seafood Takeaway serves fish and chips so good that fans have been known to sail there just to indulge. Ann Rendall's trial of beremeal battered fish and Orkney patties was a big hit, so will feature as a menu 'special' from time to time. Try Ann's recipe at home – here it is.

Makes enough batter for 6 to 8 people

350g (12oz) beremeal
1 tsp salt
Ground black pepper
2 tsp baking powder
About 900ml (1½ pt) sparkling spring water
8 fillets of fresh white fish
Beremeal to dust
Oil to deep fry

Sift the dry ingredients into a bowl and mix to a coating batter with sparkling spring water, adding more if needed. Heat the oil to 180°C/350°F. Test with a drop of batter: it will sizzle and slowly crisp when the oil is at the right temperature. Dust the fish with beremeal. Dip into the batter and fry in batches of one or two, turning to make sure the fish is evenly cooked through. This takes 4 to 5 minutes. Drain on kitchen towel. Keep warm in the oven till all the fish is cooked. Serve hot with fresh fried chips and sauces. Oh, for a Westray fish supper!

Scottish sea trout in beremeal

Graeme and Anne Finlayson were chef patrons of the Eilean Dubh restaurant at Fortrose which they sold in 2017. Now both employ their skills to create new dishes for a wider market. Here is one for you to enjoy.

Serves 2

50g (1¾oz) butter
2 cloves garlic, minced
1 small tsp dried parsley
Salt and ground black pepper
1 tsp harissa powder
50g (1¾oz) beremeal
1 egg beaten with 1 tbsp crème fraiche
185g (6oz) seatrout fillet
200g (7oz) spiced couscous
Boiling water
Rapeseed oil to fry

4 baby leeks, trimmed
4 cherry tomatoes
A handful of baby salad leaves
Watermelon, optional

Melt the butter and stir in garlic, parsley, a pinch of salt, pepper and harissa. Add the beremeal, stir together and set aside. Put the couscous in a bowl, cover with boiling water and leave to soak. Heat a little rapeseed oil in a frying pan and stir-fry the leeks and tomatoes for 2 minutes to soften. Remove and set aside. Dip the fish in the egg mixture, then roll it in the beremeal mix, pressing it onto the surface of the fish. Fry the fish in a little rapeseed oil on medium heat, skin-side down, for 4 minutes, turn and cook the top for 2 or 3 minutes till firm. Serve on a bed of couscous, garnished with salad leaves, leeks, tomato and watermelon fingers if liked.

Hand-raised North Ronaldsay mutton pie with sugar kelp jelly

Chef Sam Britten learned his craft working in Michelin-star establishments in Britain and Australia, now he has settled with his wife and young son in Orkney. In this recipe, he has put his skills to good use creating a dish around two ingredients unique to Orkney: beremeal and North Ronaldsay mutton.

Makes a pie 18cm (7in) across: serves 6 to 8

For the stock:
2 pigs' trotters

1 stick of dried sugar kelp (or dried kombu seaweed)
1 onion, chopped

For the filling:
1.1kg (2½ lb) North Ronaldsay mutton shoulder,
    diced or coarsely minced, or use mutton or hoggit
    (year-old lamb)
10g (2 tsp) salt
Ground black pepper

For the pastry:
200g (7oz) rendered mutton fat or lard
220ml (7½fl oz) water
200g (7oz) plain flour
175g (6oz) beremeal
5g (1 tsp) salt
Beaten egg

The day before, put the pigs' trotters, sugar kelp and onion in a pan, cover with cold water and bring to the boil. Reduce the heat and simmer slowly for 5 to 6 hours. Strain and chill overnight to set to a gel. The following day, heat the oven to 160ºC/fan 140ºC/325ºF/gas 3. Oil a 18cm (7in) non-stick spring-form cake tin. Season the mutton with salt and pepper and mix well. Heat the water and fat to just boiling. Sift the dry ingredients into a mixing bowl, add the hot water and fat and beat to a smooth dough. Cover till cool enough to handle. Cut off a quarter of the dough and reserve for the lid. Roll out the remaining dough to a circle and place in the base of the cake tin. While the dough is pliable, press it evenly over the base and up the sides, approximately 0.5cm (¼in) thick. Alternatively, hand-form the pastry around a pie dolly. (A wooden pie dolly is a mould round which pastry is shaped into a pie shell; the cake tin is a simple alternative.) Fill with seasoned mutton. Roll out the dough for the lid, cover the pie and pinch round the edges to seal. Stamp a small hole in the middle to allow steam to escape. Brush with beaten egg. Bake for 80 to 90 minutes till golden and firm. Cool for 15 minutes. Heat some of the pre-prepared gelled stock. Use a turkey baster to pour the stock through the top of the pie. Chill overnight to set. The pie will keep chilled for up to three days. Serve with boiled new Orkney tatties.

Westray Wife quiche

When plans went wrong at the last minute, hospitality lecturers at Orkney College, University of the Highlands and Islands, Anne Hill and Ingrid Groat, came to the rescue. They helped prepare and serve the opening buffet for the Orkney International Science Festival, 2016. On the menu was a quiche made with 'Westray Wife', a local semi-hard artisan cheese.

Makes a quiche 20cm (8in) in diameter

For the pastry:
85g (3oz) plain flour
85g (3oz) beremeal
85g (3oz) margarine
Salt and ground black pepper
Water to mix

For the filling:
175g (6oz) grated 'Westray Wife' cheese or similar
    (for example Ayrshire Dunlop or any Gouda-style
    artisan Scottish cheese)
4 tbsp thick natural yoghurt – we used Westray
    yoghurt
Milk to mix
4 large eggs, beaten
75g (3oz) spring onions, washed, trimmed and
    chopped
Salt and ground black pepper

Sift the flours into a bowl and season with salt and pepper. Rub in the margarine. Mix with cold water to a pliable dough and rest for 5 minutes. Roll out the pastry, line a flan dish and chill.

Heat the oven to 180°C/fan 160°C/350°F/gas 4. Scatter the grated cheese and spring onions over the pastry case. Beat together eggs, yoghurt and enough milk to cover the cheese, season and pour over. Bake for 25 to 30 minutes till golden. Enjoy warm or chilled.

Salmon quiche

A quick ready-to-eat meal I made for friends who had a long drive home.

Makes a quiche 20cm (8in) in diameter

For the pastry:
85g (3oz) rolled oats, blitzed to a coarse flour
85g (3oz) beremeal

¼ tsp fine sea salt
85g (3oz) butter or margarine
1 tbsp cold water
Beremeal to dust

For the filling:
45g (1½oz) finely chopped leek or spring onion
15g (½oz) butter
150g (5oz) canned sweetcorn, drained
115g (4oz) chopped smoked salmon
3 medium eggs, beaten with
150ml (¼ pt) milk
Sea salt and ground black pepper

For the topping:
15g (½oz) parmesan or other hard cheese, grated

Heat the oven to 180ºC/fan 160ºC/350ºF/gas 4. Oil the flan dish. Put the beremeal, oats and salt into a bowl and rub in the butter. Mix with water to make a pliable dough. Knead on a well-floured board and shape into a round. Carefully roll into a circle approximately 28cm (11in) diameter to line the flan dish. Line the flan with baking parchment and fill with ceramic baking beans or dried pulses. Bake for 15 minutes or till the pastry is firm. Remove the parchment and beans and bake a further 5 minutes to crisp. Make the filling. Soften the leeks in melted butter, add the sweetcorn and salmon and mix together. Spread evenly in the pastry case. Beat the eggs and milk together, season and pour over. Bake for 20 minutes, then dust with grated cheese. Bake for a further 10 minutes till golden. Serve hot with salad, or it is equally good chilled.

Leek and mushroom quiche

Beremeal pastry is more difficult to handle but worth the effort.

Makes a quiche 20cm (8in) in diameter

For the pastry:
115g (4oz) beremeal
Pinch of salt
60g (2oz) margarine
Water to mix

For the filling:
45g (1½oz) leek, finely chopped
15g (4oz) sliced mushrooms
20g (¾oz) butter
115g (4oz) Orkney cheddar, grated
3 eggs, beaten
150ml (5fl oz) milk
Sea salt and ground black pepper

Heat the oven to 180ºC/fan 160ºC/350ºF/gas 4. Oil a 20cm (8in) flan dish. Sift the dry ingredients into a bowl and rub in the margarine. Mix to a stiff pliable dough with cold water and rest for 10 minutes. Knead on a beremeal-dusted board, cover with clingfilm and roll out to fit the flan dish. Remove the clingfilm, lift the pastry over the rolling pin and line the tin. The pastry may break but is easily patched. Bake blind as for salmon quiche on page 113. Make the filling. Soften the leeks and mushrooms in butter, then season with salt and ground black pepper. Beat the eggs and milk

An old-fashioned Orcadian wooden dresser

together and season. Drain the leek and mushrooms and spoon them over the base of the pastry case. Cover with grated cheese and pour the egg and milk mixture over. Bake for 20 to 25 minutes till bubbling and golden. Serve hot or cold.

Useful tips:
A pre-baked pastry case keeps up to 1 week in a sealed container.

Substitute blanched broccoli or shredded carrot for the mushrooms.

Puddings

Foveran beremeal chocolate fondants

Chef Paul Doull's first job was as pot-wash boy in the kitchen at the Foveran, St Ola. Now chef patron there, he is an example of what can be achieved by dedicated hard work and enthusiasm. A visit to the Foveran is a feast of local produce and wonderful views. This recipe is for one of their famous puddings.

Makes 8 round individual pudding basins 5cm (2in) high and 7.5cm (3in) diameter

50g (1¾oz) melted butter for brushing
Cocoa powder for dusting
200g (7oz) good quality dark chocolate, chopped into small pieces
200g (7oz) butter, in small pieces
200g (7oz) caster sugar
4 eggs and 4 egg yolks
100g (3½oz) plain flour
100g (3½oz) beremeal

First prepare the pudding basins. Using upward strokes, brush melted butter heavily over the inside of the moulds and place on a tray in the fridge or freezer to set. Brush with melted butter once more and dust each with a teaspoon of cocoa powder. Tap out any excess.

Place a bowl over a pan of barely simmering water, then slowly melt the chocolate and butter together. Remove from the heat and stir till smooth. Cool for about 10 minutes. In a separate bowl, whisk the eggs, egg yolks and sugar until thick and pale and the whisk leaves a trail in the mixture. Sift the flour and beremeal into the whisked eggs and beat together. Pour the melted chocolate into the egg mixture a third at a time, beating well after each addition till all is mixed to a loose fondant batter. Tip the batter into a jug, then divide between the moulds. The fondant can be frozen for up to one month and cooked from frozen. Chill for at least 20 minutes. This can be prepared the day before and chilled overnight. To bake from frozen simply add 5 minutes to the baking time.

Heat the oven to 200°C/fan 180°C/400°F/gas 6. Place the fondants on a baking tray. Bake for 10 to 12 minutes until the tops have formed a crust and they are starting to come away from the sides of the moulds. Remove from the oven and wait for 1 minute before turning out. Loosen the fondant tops very gently so they come away from the sides, easing them out of the moulds. Tip each fondant slightly onto your hand so you know it has not stuck and tip back into the mould ready to serve.

To serve, starting from the middle of each plate, squeeze a spiral of caramel sauce. Prepare the number of plates you need. Carefully invert a fondant onto the middle of each plate. Using a large spoon dipped in hot water, scoop a spoon of your favourite ice cream and serve at the side of each fondant.

Bere bannock and whisky ice cream

A recipe created by a friend of mine, food writer Rosemary Moon. We have much in common, including a love of Orkney.

Serves 6 to 8

450ml (15fl oz) milk
½ tsp vanilla essence
4 large egg yolks
100g (3½oz) caster sugar
300ml (10fl oz) double cream
100g (3½oz) bere bannock crumbs, dried in a slow
    oven or under the grill
3 tbsp whisky (not a peaty one)

To make the custard, heat the milk and vanilla until almost boiling. Meanwhile whisk the egg yolks and sugar until pale. Pour the milk onto the eggs, whisking all the time. Rinse the pan out with cold water (this prevents the custard from catching on the bottom). Return the mixture to the pan. Cook over a low heat, stirring continuously until the custard thickens enough to coat the back of a wooden spoon. Pour into a bowl and chill for at least 1 hour. Whisk the cream till it is thickly floppy, mix with the chilled custard, whisky and 50g to 75g of the bannock crumbs. Don't be tempted to add more whisky because it will not freeze. Freeze-churn in an ice cream machine or freeze in a suitable container, removing it from the freezer and whisking every hour for a smooth result. Serve the machine-made ice cream immediately or transfer to a suitable container and keep in the freezer. Serve the ice cream sprinkled with a few dried bannock crumbs.

Chocolate orange tart

Beremeal is good for you, and with the benefits of eating dark chocolate the two are a good match.

Makes a tart 20cm (8in) diameter

For the rich beremeal shortcrust pastry:
100g (3½oz) beremeal
100g (3½oz) plain flour
100g (3½oz) butter
15g (½oz) caster sugar
1 egg yolk

For the chocolate orange filling:
60g (2oz) dark chocolate
60g (2oz) butter
Grated rind of 1 orange
2 eggs
75g (2½oz) caster sugar
30g (1oz) beremeal
Icing sugar

Heat the oven to 180ºC/fan 160ºC/350ºF/gas 4. Sift the flours into a bowl and rub in the butter. Stir in the sugar. Mix with the egg yolk and sufficient cold water to a pliable dough. Rest for 5 minutes, then roll out the dough and line the tin. Bake blind as for the recipe on page 113. Remove the parchment and baking beans and bake for a further 5 minutes to crisp.

Melt the chocolate in a bowl over a pan of simmering water. Stir in the butter and orange rind. Remove from the heat. Whisk the eggs and sugar till thick and creamy. Fold in the sifted beremeal. Stir in the chocolate mixture and pour into the baked flan case. Bake for 15 to 20 minutes. Test by inserting a wooden cocktail stick, which should be slightly sticky when removed. Do not over-bake. Cool. Dust with icing sugar. Serve warm or cold.

Toffee apple pudding

In this pudding a rich toffee apple sauce lies hidden below light sponge.

Serves 3 or 4: use a 600ml (1pt) pie dish

1 large cooking apple, peeled, cored and sliced.
45g (1½oz) self-raising flour
45g (1½oz) beremeal
2 tsp baking powder
50g (1¾oz) caster sugar
45g (½oz) melted butter
1 egg, beaten
100ml (3½fl oz) milk
1 tsp vanilla essence

For the topping:

75g (2½oz) dark soft brown sugar
120ml (4fl oz) boiling water

Heat the oven to 180ºC/fan 160ºC/350ºF/gas 4. Butter the dish. Lay overlapping apple slices over the base. Sift the flours and baking powder into a bowl and add the sugar. Beat the butter, egg, milk and vanilla essence together. Pour into the flour mix, stir together and pour over the apples. Mix the sugar and water and pour over the top. Bake for 30 minutes till risen and set. Dust with caster sugar. Serve warm with cream or ice cream.

Cranachan Orkney style

This is another recipe from Orcadian food writer Alan Bichan. 'I've altered the usual cranachan recipe to include Highland Park and a little beremeal. Not only does this "Orcandise" the recipe, but it also produces a delightful and unusual finale to a Burns supper.'

Serves 4

25g (scant 1oz) medium oatmeal
15g (½oz) beremeal
300ml (½ pt) whipping cream
25g (scant 1oz) caster sugar
2 tbsp Highland Park whisky
275g (10oz) raspberries

Put the oatmeal and beremeal in a saucepan and cook over a gentle heat, stirring frequently. The mixture will emit some harmless fumes, so you may want to open a window before you start! When a darker brown colour is achieved, remove from the heat and cool on a plate. Whip the cream and stir in the sugar, whisky and cooled oatmeal mixture. Divide the mixture between four glasses, layered with the raspberries. Serve with thin shortbread if liked.

Brewing

Home-brewed ale

In *Under Brinkie's Brae* (1979) George Mackay Brown writes: 'Old islanders used to say that the ale brewed in March is best of all. In a Birsay house, last week, I sipped a glass made from corn-malt. It was certainly the most delectable ale I have tasted this year. I sipped, and swilled and swallowed: and felt suddenly without a care in the marvellous rich beautiful world. I don't know whether it was March ale or not, but Mrs Matches the brewer was right when she said it would be worth coming from Stromness to experience. No brewery puts out such marvellous beer, believe me.'

How to make home-brewed 'corn ale'

An original recipe from Warsetter, Sanday, dated 1930.

Making malt

Malted bere (corn) is darker than that made by malting barley. For malt the corn should be of good quality, not too snug [tightly packed together]. First steep in water for 24 to 30 hours, then strain off the water. The corn should be spread out on a warm loft about 4 to 5in deep at first, then a little deeper as the malt dries. It should be turned every day to ensure regular growth and a little water sprinkled over it. When turning, care should be taken to see that the outside edges are put to the inside to ensure regular growth. When the sprouts are ¼ in long the malt has to be heaped together in a close mound, and covered with sacks until it takes heat and should be heated for 30 to 36 hours. This is called, in 'sweetbed'. The outside of the heap should be put to the inside at half time to ensure regular heating. After the malt has been in 'sweetbed' it should be spread out to cool before going to the mill to be bruised and put on the kiln without delay.

Brewing the ale

1 stone (14lb) of malt makes 4 gallons of ale. Put the malt into a tub, cover with approximately 4 or 5 gallons of warm water at 75°C (167°F), until it becomes the consistency of very thin porridge. Let it stand till all the good is out of the malt, about 2 hours. Then strain it, add 2oz or so of hops and boil for 1 hour. A little sugar may be added at this stage to increase the strength of the brew. Not too much, it will kill the yeast. Strain and cool to blood heat. It is time to 'barm'. Add a generous ½ cup of barm or 1oz (30g) brewer's yeast. The barm should be put in a pail and some ale added now and again till the pail is ¾ full. Then pour it all into the ale, which should be in a large enough tub to allow it to froth in barming (fermenting). The barm will work in a few hours and should do so for a day or two depending on the weather. In spring 3 to 5 days, but in winter it takes longer. When it stops it is time to bottle the ale in sterilised bottles. Keep at least 1 month before drinking. Some add ½ tsp sugar to each bottle as they fill to sweeten the ale.

(Barm is the yeasty sediment left at the bottom of a barrel of brewed ale. It was saved to start the next brew and often shared between brewers.) The miller always knew who was brewing because he kilned the malt! Barm keeps in a cool place and may be frozen. Rae Phillips' father, miller at the Barony Mill, once put barm in a sealed jar in the mill burn to keep, but it was washed away in the current!

Not far from the Barony Mill two cottages stood side by side, 'Hell' and 'Purgatory'. George Mackay Brown's aunt, Annie Johnston, a renowned brewer, resided in 'Hell'. George Mackay Brown wrote about one of his visits: 'And each evening, before I went to bed, my hostess heated a bowl of ale for me, which led to quick felicitous dreamless sleep . . . Not in any bar on earth is it possible to buy ale of such quality.' (*Under Brinkie's Brae*, 1979.)

> 'Here's health tae ye and yers
> For being so kind to we and wiz;
> And if ever ye and yirs come to we and wiz,
> We and wiz sal be as kind to ye and yirs
> As ever ye and yirs was to we and wiz.'
> *North Ronaldsay man's toast.*

Acknowledgements

I owe a great debt of gratitude to all those who have helped me to compile this book.

In particular I would like to thank retired miller Rae Phillips, who has patiently guided me along the way, and artist Ruth Tait whose illustrations make this book truly 'Orkney'. Both are original members of the Birsay Heritage Trust.

Thank you also to archivist Lucy Gibbon and Margaret Rendall who searched old documents on my behalf. Also to archaeologist Caroline Wickham-Jones who advised on the origins of bere and to proof-readers Valerie and Kathryn.

It is not possible to mention everyone in person but suffice to say I have been greatly supported by members of the Birsay Heritage Trust, Orkney Family History Society, Radio Orkney, Orkney College UHI, food writers and through conversation with experts.

A big thank you to all who contributed such stunning recipes: friends, family, Orcadians local and further afield, professional and home cooks, chefs and food writers, shops, cafes and hotels. Not forgetting all who tried, tested and tasted.

An old broken millstone, quarried from millstone grit and once used to mill bere at the Barony Mill

For Further Reading

Archives at Kirkwall Library, Orkney Heritage Society Fereday Project, and Eric Linklater's notes on Orkney history and folklore.

George Mackay Brown, *Under Brinkie's Brae* (1979)

H. W. M. Cant and H. N. Firth, *Light in the North: St Magnus Cathedral through the Centuries* (1990)

Alexander Fenton, *The Northern Isles: Orkney and Shetland* (1978)

John Firth, *Reminiscences of an Orkney Parish* (1974)

Margaret Flaws and Gregor Lamb, *The Orkney Dictionary* (from *The Orcadian*, Kirkwall)

Enid Gauldie, *The Scottish Country Miller* (1981)

F. Marian McNeill, *The Scots Kitchen, Its Traditions and Lore* (1929, new edition 2010)

William P. L. Thomson, *The New History of Orkney* (third edition, 2008)

David M. N. Tinch, *Shoal and Sheaf: Orkney's Pictorial Heritage* (1989)

Caroline Wickham-Jones, *Fear of Farming* (2010)

The website of the James Hutton Institute has regular updates on the progress of bere research: *http://www.hutton.ac.uk/news/understanding-living-heritage-bere-barley-more-sustainable-future*

Where and How to Buy Bere and Beremeal

Contact Barony Mill to order direct or to find out your nearest stockist.

Website: *www.birsay.org.uk/baronymill.htm*

Tel: 01856 721439 or 01856 721309

Email: miller@birsay.org.uk

The mill cat enjoys a snooze